DON'T TRUST
YOUR GUT

DON'T TRUST YOUR GUT

YOUR GUT

USING DATA TO GET WHAT YOU REALLY WANT IN LIFE

SETH STEPHENS-DAVIDOWITZ

DEY ST.

An Imprint of WILLIAM MORROW

HarperCollins books may be purchased for educational, business,
or sales promotional use. For information, please email the Special
Markets Department at SPsales@harpercollins.com.

FIRST EDITION

Library of Congress Cataloging-in-Publication Data has been applied for.

ISBN 978-0-06-288091-8 (hardcover)
ISBN 978-0-06-323937-1 (international edition)

22 23 24 25 26   LSC   10 9 8 7 6 5 4 3 2 1

CONTENTS

INTRODUCTION: SELF-HELP FOR DATA GEEKS

You can make better life decisions. Big Data can help you.

We are living through a quiet revolution in our understanding of the most important areas of human life—thanks to the internet and all the data it has created. In the past few years, scholars have mined a variety of enormous datasets—everything from OkCupid messages to Wikipedia profiles to Facebook relationship statuses. In these thousands or millions of data points, they have found, for perhaps the first time, credible answers to fundamental questions. Questions such as:

- » What makes a good parent?
- » Who is secretly rich—and why?
- » What are the odds of becoming a celebrity?
- » Why are some people unusually lucky?
- » What predicts a happy marriage?
- » What, more generally, makes people happy?

Often, the answers revealed in the data are not what you might have guessed, and they suggest making different decisions than you might otherwise make. Quite simply, there are insights in these mounds of new data that can allow you, or someone you know, to make better decisions.

Here are three examples uncovered from researchers studying very different parts of life.

Example # 1: Suppose you are a single man or woman who isn't getting as many dates as you would like. You try to improve yourself in every way that others suggest. You dress better. You whiten your teeth. You get a pricey new haircut. But still. The dates, they're not coming.

Insights from Big Data might help.

The mathematician and author Christian Rudder studied tens of millions of preferences on OkCupid to learn the qualities of the site's most successful daters. He found—and this was not at all surprising—that the most prized daters are those blessed with conventional beauty: the Brad Pitts and Natalie Portmans of the world.

But he found, in the mounds of data, other daters who did surprisingly well: those with extreme looks. Think, for example, of people with blue hair, body art, wild glasses, or shaved heads.

Why? The key to these unconventional daters' success is that, while many people aren't especially attracted to them, or find them plainly unattractive, some people are *really* attracted to them. And in dating that is what is most important.

In dating, unless you are drop-dead gorgeous, the best strategy is, in Rudder's words, to get "lots of Yes, lots of No, but very little Meh." Such a strategy, Rudder discovered, can lead to about 70 percent more messages. Be an extreme version of yourself, the data says, and some people will find you extremely attractive.

And example # 2: Suppose you just had a baby.* You need to pick a neighborhood in which to raise this child. You know the drill. You consult a few friends, Google some basic facts, visit a couple of homes. And voila! You've got yourself a home for your family. You assume there isn't much more of a science to this.

There is a science to neighborhood-hunting now.

Researchers recently took advantage of newly digitized tax records to study the life trajectories of hundreds of millions of Americans. The scientists discovered that being raised in certain cities—and even certain blocks within those cities—can dramatically improve a person's life outcomes. And these great neighborhoods are not necessarily the ones people suspect. Nor are they the ones that cost the most. There are now maps that can inform parents, based on extensive data analysis, about the quality of every tiny neighborhood of the United States.

That's not all. Researchers have also mined data to find traits that the best neighborhoods for raising kids tend to share; in the process, they have upended much conventional wisdom about child-rearing. Thanks to Big Data, we are finally able to tell parents what really matters for raising a successful kid (hint: adult role models) and what matters a lot less (hint: the fanciest schools).

And example # 3: Suppose you are an aspiring artist who can't seem to catch your big break. You buy every book you can on your craft. You get feedback from your friends. You re-

---

* *Mazel tov!*

vise your pieces again and again and again. But nothing seems to work. You can't figure out what you are doing wrong.

Big Data has uncovered a likely mistake.

A recent study of the career trajectories of hundreds of thousands of painters, led by Samuel P. Fraiberger, has uncovered a previously hidden pattern in why some succeed, and others don't. So, what's the secret that differentiates the big names from the anonymous strugglers?

It is often how they present their work. Artists who never break through, the data tells us, tend to present their work to the same few places over and over again. The artists who make it big, in contrast, present to a far wider set of places, allowing themselves to stumble upon a big break.

Many people have talked about the importance in your career of showing up. But data scientists have found it's about showing up to a wide range of places.

This book isn't meant to give advice only for single people, new parents, or aspiring artists—though there will be more lessons here for all of them. My goal is to offer some lessons in new, big datasets that are useful for you, no matter what stage of life you are in. There will be lessons recently uncovered by data scientists in how to be happier, look better, advance your career, and much more. And the idea for the book all came to me one evening while . . . I was watching a baseball game.

MONEYBALL FOR YOUR LIFE

I and other baseball fans can't help but notice: baseball is a very different game than it was three decades ago. When I was a young boy and cheering on my beloved New York Mets, baseball teams made decisions using their gut and intuition. They chose whether to bunt or steal based on the feelings of the manager. They chose which players to draft based on the impressions of scouts.

However, in the latter part of the twentieth century, there were hints of a better way. Every year of my childhood, my father would bring home a new book by Bill James. James, who worked the night shift as a security guard at a pork and beans cannery in Kansas, was an obsessive fan of baseball. And he had a nonstandard approach to analyzing the game: newly available computers and digitized data. James and his peers—called sabermetricians—discovered, in their data analysis, that many of the decisions that teams typically made, when they relied on their gut, were dead wrong.

How much should teams bunt? Much less, the sabermetricians said. How much should they steal? Almost never. How much were players who drew a lot of walks worth? More than teams thought. Whom should teams draft? More college pitchers.

My father wasn't the only one intrigued by James's work. Billy Beane, a baseball player turned baseball executive, was a big Bill James fan. And, when he became general manager of the Oakland A's, he chose to run his team using the principles of sabermetrics.

The results were remarkable. As famously told in the book and movie *Moneyball*, the Oakland A's, despite having one of the lowest payrolls in baseball, reached the playoffs in 2002 and 2003. And the role of analytics in baseball has exploded since then. The Tampa Bay Rays, who have been called "a team more Moneyball than the Moneyball A's themselves," reached the 2020 World Series despite the third-lowest payroll in baseball.

Further, the principles of Moneyball and the powerful underlying idea—that data can be useful in correcting our biases—have transformed many other institutions. Other sports, for example. NBA teams increasingly rely on analytics that track the trajectory of every shot. In data from 300 million shots, large deviations from optimal shooting have been found. The average NBA jump shooter, it turns out, is twice as likely to miss a shot too short as opposed to too long. And, when he shoots from the corner, he is more likely to miss to the side away from the backboard, perhaps because he is too afraid of hitting it. Players have utilized such information to correct these biases—and make more shots.

Silicon Valley firms have been built largely on Moneyball principles. Google, where I formerly worked as a data scientist, certainly believes in the power of data to make major decisions. A designer famously quit the company because it frequently ignored the intuition of trained designers in favor of data. The final straw for the designer was an experiment that tested forty-one shades of blue in an ad link on Gmail to collect data on which one would lead to the most clicks. The designer may have been frustrated, but the data experiment netted Google an estimated $200 million per year in additional ad revenue and Google has never wavered on its belief in data as it built its $1.8 trillion company. As Eric Schmidt, its former CEO, put it, "In God we trust. All others have to bring data."

James Simons, a world-class mathematician and founder

of Renaissance Technologies, brought rigorous data analysis to Wall Street. He and a team of quants built an unprecedented dataset of stock prices and real-world events and mined it for patterns. What tends to happen to stocks after earnings announcements? Bread shortages? Company mentions in newspapers?

Since its founding, Renaissance's flagship Medallion fund—trading entirely based on patterns in data—has returned 66 percent per year before fees. Over the same time period, the S&P 500 has returned 10 percent per year. Kenneth French, an economist associated with the efficient market hypothesis, which suggests it is virtually impossible to meaningfully outperform the S&P 500, explained Renaissance's success as follows: "It appears that they're just better than the rest of us."

But how do we make big decisions in our personal lives? How do we pick whom to marry, how to date, how to spend our time, whether to take a job?

Are we more like the A's in 2002 or the other baseball teams back then? More like Google or a mom-and-pop shop? More like Renaissance Technologies or a traditional money manager?

I would argue that the vast majority of us, the vast majority of the time, rely heavily on our gut to make our biggest decisions. We might consult some friends, family members, or self-proclaimed life gurus. We might read some advice that isn't based on much. We might squint at some very basic stats. Then, we will do what feels right.

What might happen, I wondered, as I watched that baseball game, if we took a data-based approach to our biggest life decisions? What if we ran our personal lives the way that Billy Beane ran the Oakland A's?

I knew that such an approach to life is increasingly possible these days. My previous book, *Everybody Lies*, explored how all the new data available thanks to the internet is transforming our understanding of society and the human mind. The stats revolution may have come to baseball first thanks to all the data that its stats-obsessed fans had demanded and collected. The Lifeball revolution is now possible thanks to all the data that our smartphones and computers have collected.

Consider this not-too-trivial question: what makes people happy?

Data to answer this question in a rigorous, systematic way was not available in the twentieth century. When the Moneyball revolution hit baseball, sabermetricians may have been able to analyze data from play-by-plays that had been dutifully recorded for every game. Back then, however, data scientists did not have the equivalent of play-by-plays for people's life decisions and resulting mood. Back then, happiness, unlike baseball, was not open to rigorous quantitative research.

But it is now.

Brilliant researchers such as George MacKerron and Susana Mourato have utilized iPhones to build an unprecedented happiness dataset—a project they call Mappiness. They recruited tens of thousands of users and pinged them on their smartphones throughout the day. They asked simple questions such as what they were doing, who they were with, and how happy they were. From this they created a dataset of more than 3 million happiness points, a far cry from the dozens of data points that had previously been the stuff of happiness research.

Sometimes the results in these millions of data points are provocative, such as that sports fans get more pain from their teams' losses than they gain pleasure from their teams' wins. Sometimes the results are counterintuitive, such as that drinking alcohol tends to give a bigger happiness boost to someone doing chores than someone socializing with friends. Sometimes the results are profound, such as that work tends to make people miserable unless they work with their friends.

But, always, the results are useful. Ever wondered precisely how weather affects our mood? Which activities tend to systematically deceive us in how much pleasure they will bring? The real role that money plays in happiness? How much our surroundings determine how we feel? We now, thanks to MacKerron, Mourato, and others, have credible answers to all these questions—answers that will be the stuff of Chapters 8 and 9. In fact, I will even conclude this book with a reliable formula for happiness that has been uncovered in millions of smartphone pings. I call it the Data-Driven Answer to Life.

So, for the past four years, motivated by a baseball game, I have disappeared into intensive study. I have talked to researchers. I have read academic papers. I've pored over the appendices of papers in ways that I am pretty sure no researcher was expecting. And I've done some of my own research and interpretation. I viewed my job as finding the Bill Jameses of arenas such as marriage, parenting, athletic achievement, wealth, entrepreneurship, luck, style, and happiness—and allowing all of you to become the Billy Beane of your personal lives. I am now ready to report everything that I have learned.

Call it "Moneyball for your life."

THE INFIELD SHIFTS OF LIFE

Before I explored the research, I asked myself some basic questions. What might a life built on Moneyball principles look like? How might our personal decision-making look if, like the A's and the Rays, we followed the data instead of our instincts? One thing that is striking from watching baseball post-*Moneyball* is that some of the decisions made by analytics-driven baseball teams seem . . . well, a little odd. Consider this example: the location of infielders.

In the post-*Moneyball* era, baseball teams increasingly engage in infield shifts. They load up many of their defenders in the same part of the field, leaving wide swaths of the field completely unguarded, seemingly wide-open for a hitter to direct the ball. The infield shift looks positively insane to fans of traditional baseball. But insane it is not. Such shifts are justified in mounds of data that predict where particular players are most likely to hit the ball. The numbers tell baseball teams that, even though it looks wrong, it is right.

If we take a Moneyball approach to life, we might similarly expect to find that some seemingly odd decisions—call them the infield shifts of life—are justified.

We've already discussed a couple. Shaving your head or dyeing your hair blue to get more dates is an infield shift of life. Here's another one, uncovered in the Big Data of sales.

Suppose you are trying to sell something. This is an increasingly common experience. As the author Daniel Pink put it in his book *To Sell Is Human*, whether we are "pitching colleagues, persuading funders, [or] cajoling kids . . . we're all in sales now."

Anyway, whatever your pitch, you give it your best shot.

You write up your pitch. (This is good!) You practice your pitch. (Good!) You get a good night's sleep. (Good!) You eat a hearty breakfast. (Good!) You fight through your nerves and get up there. (Good!)

And, as you make your sales case, you remember to convey your excitement with a big, hearty, toothy smile. (This is . . . not good.)

A recent study analyzed the effects of a salesperson's emotional expression on how much they sell.

The dataset: 99,451 sales pitches on a livestreaming retailing platform. (These days, people are increasingly buying products on services such as Amazon Live, which allows people to pitch their products by video to potential customers.) Researchers were given videos of each sales pitch along with data on how much product was sold afterward. (They also had data on the product being sold, the price of the product, and whether they offered free shipping.)

The methods: artificial intelligence and deep learning. The researchers converted their 62.32 million frames of video into data. In particular, the AI was able to code the emotional expression of the salesperson during the video. Did the salesperson appear angry? Disgusted? Scared? Surprised? Sad? Or happy?

The result: the researchers found that the emotional expression of a salesperson was a major predictor of how much product they sold. Not surprisingly, when a salesperson expressed negative emotions, such as anger or disgust, they sold less. Rage doesn't sell. More surprisingly, when a salesperson expressed highly positive emotions, such as happiness or surprise, they sold less. Joy doesn't sell. When it comes to increasing sales, a salesperson's limiting their excitement—having a poker face instead of a smile—proves about twice as valuable as free shipping.

Sometimes, to sell your product, you should convey less enthusiasm for your product. It might feel wrong, but the data says that it's right.

FROM *EVERYBODY LIES* TO *DON'T TRUST YOUR GUT*

Brief pause while I justify this book to readers of my first book, *Everybody Lies*. Some of you may have been drawn to this book because you were fans of that book. And if that doesn't explain how you came to this book at all, perhaps I can convince you in the next few paragraphs to buy that book as well. I try.

In *Everybody Lies*, I discussed my research on how we can use Google searches to uncover what people really think and do. I called Google searches "digital truth serum" because people are so honest to the search engine. And I called Google searches the most important dataset ever collected on the human psyche.

I showed that:

» Racist Google searches predicted where Barack Obama underperformed in the 2008 and 2012 presidential elections.
» People frequently type full sentences into Google, things like "I hate my boss," "I'm drunk," or "I love my girlfriend's boobs."
» The top Google search that starts "my husband wants . . ." in India is "my husband wants me to

breastfeed him." In India, there are almost as many Google searches looking for advice on how to breastfeed a husband as there are on how to breastfeed a baby.

» Google searches for do-it-yourself abortions are almost perfectly concentrated in parts of the United States where it is hard to get a legal abortion.

» Men make more searches for information on how to make their penis bigger than how to tune a guitar, change a tire, or make an omelet. One of their top Googled questions about their penis is "How big is my penis?"

At the end of that book, I suggested my next book would be called *Everybody (Still) Lies* and would keep exploring what Google searches tell us. Sorry, I guess I lied about that. Figures, from the author of *Everybody Lies.*

This book is, on its surface, very different. And, if you were hoping to get further analysis of men's searches about

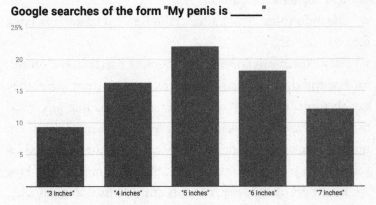

Google searches of the form "My penis is _____"

Chart: Don't Trust Your Gut by Seth Stephens-Davidowitz • Source: Google Trends • Created with Datawrapper

their genitals, you will be sorely disappointed. Eh, fine. I'll give you one more. Did you know that men sometimes type into Google full sentences stating the size of their penis? They type into Google, for example, "My penis is 5 inches." And, if you examine the data on all these searches, they reveal a close-to-normal distribution of reported-to-Google penis sizes centered around five inches.

But let's move on from my research into the wacky world of Google search data, which, as mentioned, you can learn more about in *Everybody Lies*.

Most of the studies featured in this book, unlike in *Everybody Lies*, are from other people, not from me. This book is more practical, tightly focused around self-improvement rather than explorations of random parts of modern life. Further, this book has noticeably less emphasis on sex than my previous book. Any discussion of sex in this book will not focus on the secret sexual desires or insecurities of people, topics that are heavily featured in my previous book. The discussion of sex here, instead, is limited to the question of whether sex makes people happy (spoiler: yes).

But I do think this is a natural follow-up to my first book for two reasons.

First, the motivation of this book is partly based on following the data of what readers really want, not what they say they want. After I wrote *Everybody Lies*, like any good market researcher, I asked readers what resonated with them most. Most people told me they were particularly moved by some of the sections on the world's biggest problems and how we might fix them—sections on child abuse or inequality, for example.

But, as the author of *Everybody Lies*, I was skeptical of what people said and wanted to see some other data—perhaps some digital truth serum. I looked at the most underlined sections on Amazon Kindle versions of the book. I noted that people frequently underlined passages about how they could improve their lives and rarely underlined passages about how to improve the world. People are drawn to self-help, I concluded, whether they admit it or not.

A more extensive study of Amazon Kindle data came to a similar conclusion. Researchers found, over a large sample of books, that the word "you" was twelve times more likely to appear in the most underlined sentences than other sentences. People, in other words, really like sentences that include the word "you."

Hence, the first paragraph of *Don't Trust Your Gut*:

"*You* can make better life decisions. Big Data can help *you*."

That was a data-driven, not a gut-driven, first paragraph. It was delivered to you in a book written to help you get more of what you want in your life. Did you like it?

The popularity of books that can offer help to readers is also confirmed by a deep dive into the most popular books in history. I examined the best-selling books of all time. The biggest category of nonfiction best-sellers is self-help (making up about 42 percent of the most popular nonfiction books of all time). Next biggest is memoirs of celebrities (28 percent). And third is sex studies (8 percent).

What I'm trying to say is that, by following the data, I will write this self-help book first. Then I will write *Sex: The Data*. Then I hope that will make me famous enough to write *Seth:*

Memoir of the Author Who Got Famous by Following the Data on What Books Sell.

The second connection between *Everybody Lies* and *Don't Trust Your Gut* is that this book is also about using data to uncover the secrets of modern life. One of the reasons that data is so useful in making better decisions is that basic facts about the world are hidden from us. There are secrets about who gets what they want in life that are uncovered by Big Data.

Take this secret: who is rich? Clearly, knowing this would help any person who wants to earn more money. But knowing this is complicated by the fact that many rich people don't want other people to know that they are rich.

A recent study utilized newly digitized tax records to perform, by far, the most comprehensive study of rich people. The researchers learned that the typical rich American is not a tech tycoon, corporate bigwig, or some of the other people you might naturally have expected. The typical rich American is, in the words of the authors, the owner of "a regional business," such as "an auto dealer [or] beverage distributor." Who knew?!? In Chapter 4, we will talk about why that is—and what it implies for how to pick a career.

The media also lies to us—or at least gives us a misleading impression of how the world works by only selecting certain stories to tell us. Using data to cut through those lies can lead to information that is helpful in making decisions.

An example: age and entrepreneurial success. Data has uncovered that the media gives us a distorted view of the age of typical entrepreneurs. A recent study found that the median age of entrepreneurs featured in business magazines is twenty-seven years old. The media loves telling us the sexy stories of the wunderkinds who created major companies.

But how old is the typical entrepreneur, really? A recent study of the entire universe of entrepreneurs found that the average successful entrepreneur is forty-two years old. And the odds of starting a successful business increase up until the age of sixty. Further, the advantage of age in entrepreneurship is true even in tech, a field that most people believe requires youth to master the new tools.

Surely, the advantage of age in all types of entrepreneurship is useful information for someone who has hit middle age and thinks the chance of starting a business has passed them by. In Chapter 5, we will bust a few myths about entrepreneurial success and discuss a reliable formula uncovered in data that is likely to maximize anyone's chances of creating a successful business.

When you know the data on how the world really works—and avoid the lies of individuals and the media—you are prepared to make better life decisions.

FROM GOD TO FEELINGS TO DATA

In the final chapter of *Homo Deus*, Yuval Noah Harari writes that we are going through a "tremendous religious revolution, the like of which has not been seen since the eighteenth century." The new religion, Harari says, is Dataism, or faith in data.

How did we get here?

For much of human history, of course, the most learned

people in the world placed the highest authority in God. Harari writes, "When people didn't know whom to marry, what career to choose or whether to start a war, they read the Bible and followed its counsel."

The humanist revolution, which Harari places in the eighteenth century, questioned the God-centered worldview. Scholars such as Voltaire, John Locke, and my favorite philosopher, David Hume, suggested that God was a figment of human imagination and the rules of the Bible were faulty. With no external authority to guide us, philosophers suggested that human beings guide themselves. The ways to make big decisions, in the age of humanism, Harari writes, were "listen[ing] to yourself," "watching sunsets," "keeping a private diary," and "having heart-to-heart talks with a good friend."

The Dataist revolution, which has just started and, Harari says, may take decades or more to be fully embraced, questioned the feelings-centered worldview of the humanists. The quasi-religious status of our feelings was called into question by life scientists and biologists. They discovered that, in Harari's words, "organisms are algorithms" and feelings merely "processes of biochemical calculations."

Further, legendary behavioral scientists, such as Amos Tversky and Daniel Kahneman, discovered that our feelings frequently lead us astray. The mind, Tversky and Kahneman told us, is riddled with biases.

Think your gut is a reliable guide? Not so, they said. We are frequently too optimistic; overestimate the prevalence of easily remembered stories; latch on to information that fits what we want to believe; wrongly conclude that we can explain events that, at the time, were unpredictable; and on and on and on.

"Listening to yourself" may have sounded liberating and romantic to the humanists. But "listening to yourself" sounds, frankly, dangerous after reading the latest issue of *Psychological Review* or Wikipedia's wonderful article, "List of cognitive biases."

Finally, the Big Data revolution offers us an alternative to listening to ourselves. While our intuitions—and the counsel of our fellow human beings—may have seemed to the humanists like the only sources of wisdom that we could lean on in a godless universe, data scientists are now building and analyzing enormous datasets that can free us from the biases of our own minds.

More Harari: "In the twenty-first century, feelings are no longer the best algorithms in the world. We are developing superior algorithms that utilize unprecedented computing power and giant databases." Under Dataism, "When you contemplate whom to marry, which career to pursue and whether to start a war," the answer is now "algorithms [that] know us better than we know ourselves."

I'm not quite arrogant enough to claim that *Don't Trust Your Gut* is the bible of Dataism or to try to write the Ten Commandments of Dataism, though I would love it if you thought of the other researchers whose work I discuss as the prophets of Dataism. (They really are that trailblazing.)

But I do hope that this book will show you what the new worldview of Dataism looks like, along with offering you some algorithms that might be useful to you or a friend facing a big decision. *Don't Trust Your Gut* includes nine chapters; each one explores what data can tell us about a major area of life. And the first one will focus on perhaps life's biggest decision and the decision that Harari lists first as one that might be transformed by Dataism.

So, Dataists and potential Dataism converts: can an algorithm help you pick "whom to marry"?

CHAPTER 1

THE AI MARRIAGE

Whom should you marry?

This may be the most consequential decision of a person's life. The billionaire investor Warren Buffett certainly thinks so. He calls whom you marry "the most important decision that you make."

And yet people have rarely turned to science for help with this all-important decision. Truth be told, science has had little help to offer.

Scholars of relationship science have been trying to find answers. But it has proven difficult and expensive to recruit large samples of couples. The studies in this field tended to rely on tiny samples, and different studies often showed conflicting results. In 2007, the distinguished scholar Harry Reis of the University of Rochester compared the field of relationship science to an adolescent: "sprawling, at times unruly, and perhaps more mysterious than we might wish."

But a few years ago, a young, energetic, uber-curious, and brilliant Canadian scientist, Samantha Joel, aimed to change that. Joel, like so many in her field, was interested in what predicts successful relationships. But she had a noticeably different approach from others. Joel did not merely recruit a new, tiny sample of couples. Instead, she joined together data from other, already-existing studies. Joel reasoned that, if she could merge data from the existing small studies, she could have a large dataset—and have enough data to reliably find what predicts relationship success and what does not.

Joel's plan worked. She recruited every professor she could find who had collected data on relationships—her team ended up including eighty-five other scientists—and was able to build a dataset of 11,196 couples.*

The size of the dataset was impressive. So was the information contained in it.

For each couple, Joel and her team of researchers had measures of how happy each partner reported being in their relationship. And they had data on just about anything you could think to measure about the two people in that relationship.

The researchers had data on:

» demographics (e.g., age, education, income, and race)
» physical appearance (e.g., How attractive did other people rate each partner?)
» sexual tastes (e.g., How frequently did each partner want sex? How freaky did they want that sex to be?)
» interests and hobbies
» mental and physical health

---

* *The study focused on heterosexual relationships. Further research might explore any differences among gay couples.*

» values (e.g., their views on politics, relationships, and child-rearing)
» and much, much more.

Further, Joel and her team didn't just have more data than others in the field. They had better statistical methods. Joel and some of the other researchers had mastered machine learning, a subset of artificial intelligence that allows contemporary scholars to detect subtle patterns in large mounds of data. One might call Joel's project the AI Marriage, as it was among the first studies to utilize these advanced techniques to try to predict relationship happiness.

If you like guessing games, you can try to predict the results. What do you think are the biggest predictors of relationship success? Are common interests more important than common values? How important is sexual compatibility in the long term? Does coming from a similar background as a mate make you happier?

After building her team and collecting and analyzing the data, Joel was ready to present the results—results of likely the most exciting project in the history of relationship science.

Joel scheduled a talk in October 2019 at the University of Waterloo in Canada with the straightforward title: "Can we help people pick better romantic partners?"

So, can Samantha Joel—teaming up with eighty-five of the world's most renowned scientists, combining data from forty-three studies, mining hundreds of variables collected from more than ten thousand couples, and utilizing state-of-the-art machine learning models—help people pick better romantic partners?

No.

The number one—and most surprising—lesson in the data, Samantha Joel told me in a Zoom interview, is "how unpredictable relationships seem to be." Joel and her coauthors found that the demographics, preferences, and values of two people had surprisingly little power in predicting whether those two people were happy in a romantic relationship.

And there you have it, folks. Artificial intelligence can now:

» defeat the world's most talented humans at chess and Go;
» reliably predict social unrest five days before it happens merely based on chatter on the internet; and
» inform people of an emerging health issue, such as Parkinson's disease, based on the odors they emit.

But ask AI to figure out whether a set of two human beings can build a happy life together. And it is just as clueless as the rest of us.

WELL . . . THAT SURE SEEMS LIKE A LETDOWN—AS WELL AS A truly horrific start to a chapter in my book with the bold thesis that data science can revolutionize how we make life decisions. Does data science really have nothing to offer us in picking a romantic partner, perhaps the most important decision that we will face in life?

Not quite. In truth, there are important lessons in Joel and her coauthors' machine learning project, even if computers' ability to predict romantic success is worse than many of us might have guessed.

For one, while Joel and her team found that the power of all the variables that they had collected to predict a couple's happiness was surprisingly small, they did find a few variables in a mate that at least slightly increase the odds you will be happy with them. More important, the surprising difficulty in predicting romantic success has counterintuitive implications for how we should pick romantic partners.

Think about it. Many people certainly believe that many of the variables that Joel and her team studied are important in picking a romantic partner. They compete ferociously for partners with certain traits, assuming that these traits will make them happy. If, on average, as Joel and her coauthors found, many of the traits that are most competed for in the dating market do not correlate with romantic happiness, this suggests that many people are dating wrong.

This brings us to another age-old question that has also recently been attacked with revolutionary new data: how do people pick a romantic partner?

In the past few years, other teams of researchers have mined online dating sites, combing through large, new datasets on the traits and swipes of tens of thousands of single people to determine what predicts romantic desirability. The findings from the research on romantic desirability, unlike the research on romantic happiness, has been definitive. While data scientists have found that it is surprisingly difficult to detect the qualities in romantic partners that lead to happiness, data scientists have found it strikingly easy to detect the qualities that are catnip in the dating scene.

A recent study, in fact, found that not only is it possible to predict with great accuracy whether someone will swipe left or right on a particular person on an online dating site. It is even possible to predict, with remarkable accuracy, the time it will take for someone to swipe. (People tend to take longer to swipe for someone close to their threshold of dating acceptability.)

Another way to say all this: *Good romantic partners are difficult to predict with data. Desired romantic partners are easy to predict with data. And that suggests that many of us are dating all wrong.**

WHAT PEOPLE LOOK FOR IN A PARTNER

The major development in the search for romance in the early part of the twenty-first century has been the rise of online dating. In 1990, there were six predominant ways that people met their spouses. The most frequent way was through friends, followed by: as coworkers, in bars, through family, in school, as neighbors, and in church.

In 1994, kiss.com was founded as the first modern online dating site. One year later, Match.com was started. And, in 2000, I excitedly set up my profile on JDate, an online Jewish dating site, confident that I had discovered the cool new thing . . . only to quickly realize that, once again, the cool new thing I thought I had found was actually predominantly used by weirdos like myself.

However, the use of online dating has since exploded. By

---

* *A punchier way to say this: It's easy to predict who you will* click on. *It's hard to predict who you will click* with.

How heterosexual couples have met over time

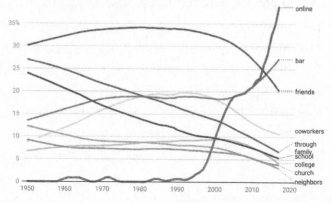

Source: Data provided by Michael Rosenfeld; first published in Rosenfeld, Thomas, & Hausen (2019) • Created with Datawrapper

2017, nearly 40 percent of couples met online. And this number continues to rise every year.

Has online dating been good for people's romantic lives? This is debatable. And many single people complain that the apps and websites lead to disappointing interactions, matches, and dates. Some recent comments on online dating on Quora, the question-and-answering website, include the following complaints: "it is exhausting"; "A significant number of the profiles of very attractive and/or flirtatious women are really Nigerian scammers"; and there are too many "unsolicited pictures of men's anatomy."

But one effect of online dating is undebatable: it has been an unambiguous gain for scientists who study romance. It is fair to say that nobody in the field of romantic science complains about the existence of dating apps and websites.

You see, in the previous century, when the courtship process happened offline, the decisions that single people made were known by a select few and forgotten shortly afterward. If scientists wanted to know what people looked for in a partner, they basically had one approach: to ask them. A groundbreaking 1947 study by Harold T. Christensen did just that. Christensen surveyed 1,157 students and asked them to rate the importance of twenty-one traits in a potential romantic partner. The number one, single most important trait reported by both men and women was "dependable character." Right near the bottom for both men and women, the traits they said they cared about least, were "good looks" and "good financial prospects."

But can we trust these self-reports? People have long been known to lie on sensitive topics. (This, in fact, was the theme of my previous book, *Everybody Lies*.) Perhaps people don't want to admit just how much they prefer to date people with pleasant faces, skinny waists, and hefty wallets.

In this century, researchers have better ways to figure out what people desire in a partner than merely asking them. When such a large percentage of courtships happens on apps or websites, daters' profiles, clicks, and messages can be retained as data. The "Yays" and "Nays" are easily coded to csv files. And researchers around the world have mined data from OkCupid, eHarmony, Match.com, Hinge, and other matching services to determine how much just about every factor contributes to one's desirability in the dating market. They have, quite simply, gathered unprecedented insights into what makes a human being desirable to other human beings.

As mentioned in the Introduction, there is some variation in what people find attractive—and daters can sometimes take advantage of this variation by occupying a niche market. However, the traits that make people more attractive, on average, are predictable.

So, what traits make people desirable to others?

Well, the first truth about what people look for in romantic partners, like so many important truths about life, was expressed by a rock star before the scientists figured it out. As Adam Duritz of the Counting Crows told us in his 1993 masterpiece "Mr. Jones": we are all looking for "something beautiful."

SOMEONE BEAUTIFUL

A team of researchers—Günter J. Hitsch, Ali Hortaçsu, and Dan Ariely—studied thousands of heterosexual users of an online dating site.

Each user of the site included photos, and the researchers recruited and paid a different group of people to rate the attractiveness of every user, based on these photos, on a scale of 1 to 10.

With the help of the 1-to-10 ratings, the researchers had a measure of conventional physical attractiveness for every dater. They could test how much looks influence how desirable someone is. They measured desirability based on how many unsolicited messages a person received and how frequently their messages were responded to.

The researchers found that looks matter. A lot.

Roughly 30 percent of how well a female heterosexual dater performed on the site could be explained by their looks. Heterosexual women are a little less shallow but still plenty

Probability that the most attractive men respond to a message from women of various looks ratings

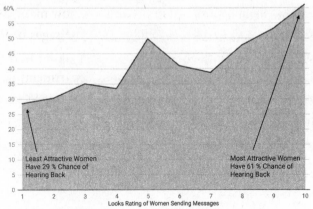

Least Attractive Women
Have 29 % Chance of
Hearing Back

Most Attractive Women
Have 61 % Chance of
Hearing Back

Looks Rating of Women Sending Messages

Source: Hitsch, Hortaçsu, and Ariely (2010); Data Provided by Günter Hirsch · Created with Datawrapper

Probability that the most attractive women respond to a message from men of various looks ratings

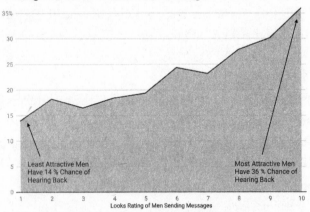

Least Attractive Men
Have 14 % Chance of
Hearing Back

Most Attractive Men
Have 35 % Chance of
Hearing Back

Looks Rating of Men Sending Messages

Source: Hitsch, Hortaçsu, and Ariely (2010); Data Provided by Günter Hirsch · Created with Datawrapper

shallow. About 18 percent of male heterosexual daters' success could be explained by their looks. Beauty, it turns out, is, for both sexes, the most important predictor of how many potential partners message and respond to one's messages in online dating.

Place that finding in the "no duh" department, as well as in the "See, I knew when people told me that looks don't matter, they were secretly superficial and thus totally and completely full of crap" department.

SOMEONE TALL (IF A MAN)

The same team of researchers that studied how looks affect daters' desirability also studied how height affected daters' desirability. (Each dater on the site reported how tall they were.)

Once again, the results were stark. A man's height had an enormous impact on how desirable he was to women. The most popular men were between 6'3" and 6'4"; such men received 65 percent more messages than men who were between 5'7" and 5'8".

The researchers also studied the effects of income on daters' desirability, which I will discuss shortly. This allowed them to make an interesting comparison between income and height in the dating market. They could ask how much more money a shorter man would have to earn to overcome his height disadvantage.

They found that a 6-foot man earning $62,500 per year is, on average, as desirable as a similar 5'6" man who earns $237,500. In other words, those six inches of height are worth about $175,000 in salary in the dating market.

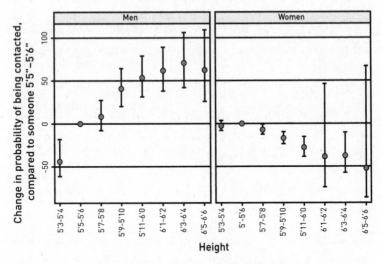

Source: *Hitsch, Hortaçsu, and Ariely (2010)*

The effect of height on desirability was reversed and far less pronounced for women. Generally, taller women had less success on the dating site. A woman who is 6'3" tall, the researchers found, could expect to receive 42 percent fewer messages than a 5'5" woman.

SOMEONE OF A DESIRED RACE (EVEN IF THEY'D NEVER ADMIT IT)

Continuing the theme of superficial factors about a person that play a disturbing role in their success on the dating market, scientists have found significant evidence of racial discrimination in dating. Christian Rudder, a mathematician who was one of the cofounders of OkCupid, analyzed data from the messages of more than one million OkCupid users. He describes the results in his fascinating book, *Dataclysm*.

The two extremely disturbing charts on the next page show the reply rates on OkCupid when heterosexual males and females of different races send messages to each other. If race did not influence dating decisions, the numbers in the chart would be exactly the same. In other words, a Black woman and a white woman sending a message to a white man would have the same chance of getting a reply. Instead, the numbers are very different. A Black woman has a 32 percent chance of getting a reply from a white male; a white woman has a 41 percent chance of getting a reply.

Overall, perhaps the most striking finding in the data is the difficulties African-American women face in the dating market. Note the second row of the first chart. Men of just about every racial group are less likely to respond to messages from Black women.

The second column of the second chart shows how African-American women respond to this harsh treatment by men: they become far less picky. For just about any group of men sending messages, Black women are the most likely to respond.

The dating experience of Black women is notably different from that of white males. White males tend to be significantly more likely to have their messages responded to. This is seen in the final row of the second chart. And they, in turn, become more picky, becoming the least likely to respond to messages from women. This is seen in the final column of the first chart.

Reply rates: female sent initial message to male

	Asian male	Black male	Hispanic male	White male
Asian female	48	55	49	41
Black female	31	37	36	32
Hispanic female	51	46	48	40
White female	48	51	47	41

Source: Data from Christian Rudder OKTrends post: https://www.gwern.net /docs/psychology/okcupid/howyourraceaffects
themessagesyouget.html • Created with Datawrapper

Reply rates: male sent initial message to female

	Asian female	Black female	Hispanic female	White female
Asian male	22	34	22	21
Black male	17	28	19	21
Hispanic male	20	31	24	22
White male	29	38	30	29

Source: Data from Christian Rudder OKTrends post: https://www.gwern.net /docs/psychology/okcupid/howyourraceaffects
themessagesyouget.html • Created with Datawrapper

Among males, the racial groups that receive the lowest response rates are Blacks and Asians.

Rudder's charts are blunt. They show the overall response rates of every racial pair, but they do not consider any other differences between the groups that might lead to differences in response rates. Perhaps some of the reason that some racial groups, on average, perform better or worse in the dating market is that some racial groups earn, on average, different incomes.

Hitsch, Hortaçsu, and Ariely try to correct for these factors. They found that, when you take into account these other factors, the bias against Asian men becomes even more severe. Since Asian men in the United States have above-average incomes, which tends to be attractive to women, the low response rates to their messages is even more striking. The researchers determined that an Asian man would have to earn a staggering $247,000 more in annual income to be as attractive to the average white woman as he would if he were white.

SOMEONE RICH

Back to findings in the no-surprise department: income affects one's desirability in dating, with the biggest effect on males.

Hitsch, Hortaçsu, and Ariely found that, all else being equal, if a man increases his earnings from $35K to $50K to $150K to $200K, he can expect the average woman to be 8.9 percent more likely to contact him. If a woman's earnings increase by the same amount, she can expect the average man to be 3.9 percent more likely to contact her.

Of course, it is well-known that a man with a substantial income can be attractive to heterosexual women. Consider the first line of Jane Austen's *Pride and Prejudice*: "It is a truth universally acknowledged, that a single man in possession of a good fortune, must be in want of a wife." Or consider the thought from the band Barenaked Ladies—who are, of course, really men: that if they "had a million dollars," they'd be able to buy someone's love.

Because the effects of wealth on romantic desirability and the efforts men make to earn more money are such cliches, I was actually surprised that the effects of income were rather modest.

Next, I will discuss the significant effects that a man's occupation has on his romantic desirability, independent of his income. For example, all else being equal, males can expect significantly more romantic attention if they are firefighters than if they are waiters.

It turns out, sometimes a switch to a different, more attractive occupation can make a male more desirable than a large salary increase. For example, the data from online dating sites suggests that a man who earned $60K in the hospitality industry would become more desirable, on average, if he earned the same amount as a firefighter than if he stayed in the same industry but upped his salary to $200K. In other words, a male firefighter who earns $60K tends to be more attractive to the average heterosexual woman than a hospitality worker who earns $200K.

While many men believe they need to earn a substantial salary to "buy" a woman's love, the data suggests that having a cool job is frequently more attractive than having a boring, but lucrative, job.

AN ENFORCER OF THE LAWS OR A HELPER OF OTHER PEOPLE IN TROUBLE (IF A MAN)

One's job matters in the mating market—if you're a man.

Hitsch, Hortaçsu, and Ariely had data from their online dating site on the occupation of daters. It turns out that a woman's occupation doesn't impact how many messages she receives, when you take into account her physical attractiveness.

For men trying to attract women, the story is different. Men who work in certain occupations receive more messages. And this is true taking into account everything else researchers know about them, including their income.

Male lawyers, police officers, firefighters, soldiers, and doctors get more messages than men who earn similar incomes, have similarly prestigious educations, are equally attractive, and are of the same height. Lawyers would be, on average, less attractive to women if they were accountants.*

---

* Seinfeld *fans may now be thinking of George Costanza. Costanza, who, according to his best friend Jerry Seinfeld, was "one of the most deceitful, duplicitous, deceptive minds of our time" and had an unstable career, frequently lied about his occupation to try to pick up women. He claimed he was a marine biologist (which presumably fits in the science/research category) and an architect (which perhaps fits best in the artist category). But data science has discovered that these occupations are not among the most attractive to women. If George wanted to lie in a data-driven way, he should have said that he was a lawyer.*

Here is the list of occupations, ranked from most to least desirable to heterosexual women on online dating sites.

DESIRABILITY OF OCCUPATIONS IN MEN (HOLDING CONSTANT INCOME)

Occupation	Percent Increase in Approaches from Women, Relative to a Student
Legal/attorney	8.6 %
Law enforcement/firefighter	7.7 %
Military	6.7 %
Health professional	5 %
Administrative/clerical/secretarial	4.9 %
Entertainment/broadcasting/film	4.2 %
Executive/managerial	4.0 %
Manufacturing	3.7 %
Financial/accounting	2.4 %
Self-employed	2.2 %
Political/government/civil	1.7 %
Artistic/musical/writer	1.7 %
Sales/marketing	1.4 %
Technical/science/engineering/research/computers	1.2 %
Transportation	1.0 %
Teacher/educator/professor	1.0 %
Student	0 %
Laborer/construction	−0.3 %
Service/hospitality/food	−3 %

Source: Hitsch, Hortaçsu, and Ariely (2010)

SOMEONE WITH A SEXY NAME

Some years ago, researchers randomly sent messages to online daters with different first names; they didn't include a photo or any other information. They found that some names were as much as twice as likely to get clicks as other names. The sexiest names (those most likely to get a response) included:

» Alexander   » Jacob
» Charlotte   » Marie
» Emma   » Max
» Hannah   » Peter

The least sexy names (those least likely to get a response) included:

» Celina   » Justin
» Chantal   » Kevin
» Dennis   » Mandy
» Jacqueline   » Marvin

SOMEONE JUST LIKE THEMSELVES

Do we look for mates who are similar to us or different than us?

Emma Pierson, a computer scientist and data scientist, studied 1 million matches on eHarmony and wrote up her results on the data journalism site FiveThirtyEight. She examined 102 traits that eHarmony measures on partners and crunched the numbers to see whether people were more likely to pair with someone who shared the trait. Pierson found it was no contest: similarity, rather than difference, leads to attraction.

Heterosexual women are especially drawn to similarity. Pierson found that, for literally every one of the 102 traits, a man's sharing the same trait was positively correlated with a woman's contacting him. This included seemingly central traits such as age, education, and income, as well as quirkier traits, such as how many photos they included in their profile or if they used the same adjectives in their profiles. A woman who describes herself as "creative" is more likely to message a man who describes himself the same way. Heterosexual men also showed a preference for women like themselves, although the preference wasn't quite as strong.*

As Pierson's FiveThirtyEight article was titled, "In the End, People May Really Just Want to Date Themselves."

Pierson's findings that similarity leads to matches was confirmed in another study, this one using Hinge data. These researchers also had a clever title for their study, "Polar Similars." The researchers also discovered a new, quirky dimension in which daters are drawn to similarity: initials. Hinge users are 11.3 percent more likely to match with someone who shares their initials. And this effect isn't driven by people from the same religions both sharing initials and matching more frequently—say, Adam Cohen matching with Ariel Cohen. The elevated match propensity of people who share the same initials holds taking into account religious affiliation.†

Opposites attract, the data tells us, is a myth. Similarity attracts—and the effects are large.

---

* For about 80 percent of the traits that Pierson studied, women showed a stronger preference for similarity than men did.
† Seinfeld fans may be thinking of Jerry Seinfeld. In the twenty-fourth episode of the seventh season, Jerry Seinfeld dates a woman, Jeannie Steinman, who is exactly like him. She doesn't merely share his initials; she, like him, has strong opinions on people's clothing decisions, orders cereal in restaurants, and says "That's a shame!" when something bad happens to a stranger.

    Jeannie S. captures the heart of Jerry S., who proposes to marry her. But he soon calls it off. "I can't be with someone like me," he concludes. "I hate myself."

    The data tells us we are all Jerry Seinfeld, looking for our Jeannie Steinman—and also may find ourselves unhappy if we find her.

WHAT PREDICTS ROMANTIC HAPPINESS

The fascinating, if sometimes disturbing, data from online dating sites tells us that single people predictably are drawn to certain qualities. But should they be drawn to these qualities?

If you are like the average single dater—predictably clicking on people with the traits the scientists found are most desired—are you going about dating correctly? Or are you dating all wrong?*

Recall, at the beginning of this chapter, I discussed the research of Samantha Joel and coauthors. Recall that they collected history's largest dataset on couples and the qualities of those couples. Recall that they found that it was surprisingly difficult to predict whether a person was happy with a romantic partner based on a large list of traits. There is not a set of traits that guarantee romantic happiness or preclude romantic happiness. And no algorithm in the world can predict, with enormous accuracy, whether two people will end up happy together.

That said, there was some predictive power in some traits, some factors that do increase the odds at least somewhat that a person will be happy in their romantic relationship. I will now discuss what *does* predict romantic happiness—and how little it has to do with the qualities that people look for in a romantic partner.

---

* *For readers looking for a thorough, science-based guide to all things romance, I recommend, in addition to Rudder's* Dataclysm, *Logan Ury's* How to Not Die Alone.

"IT'S NOT YOU, IT'S ME": THE DATA SCIENCE SAYS SO

Say there is a person, John, and he is partnered with Sally. You want to predict whether John is happy in the relationship. You are allowed to ask John and/or Sally any three questions about themselves and use this information to predict John's relationship happiness.

What questions would you want to ask? What would you want to know about the two members of this couple?

According to my read of the research of Joel and her co-authors, the best three questions to figure out whether John is happy with Sally would have nothing to do with Sally; in fact, all would be related to John. The best questions to predict John's happiness with Sally would look something like these:

» "John, were you satisfied with your life before you met Sally?"

» "John, were you free from depression before you met Sally?"

» "John, did you have a positive affect before you met Sally?"

Researchers found that people who answered "yes" to questions such as these are significantly more likely to report being happy in their romantic relationship. In other words, a person who is happy outside their relationship is far more likely to be happy inside their relationship, as well.

Further—and this was quite striking—how a person answered questions about themselves was roughly four times more predictive of their relationship happiness than all the traits of their romantic partner combined.*

Of course, the finding that one's happiness outside of a relationship can have an enormous impact on one's happiness inside that relationship is hardly a revolutionary idea. Consider this saying that was featured on Daily Inspirational Quotes: "Nobody can make you happy until you're happy with yourself first."

This is the type of quote that often makes cynical data geeks like myself roll our eyes. However, now, after reading the work of Joel and her coauthors, I have become convinced that this quote is largely true.

This relates to an important point about living a data-driven life. We data geeks may be most excited when we learn of a finding that goes against conventional wisdom or cliched advice. This plays to our natural need to know something that the rest of the world doesn't. But we data geeks must also accept when the data confirms conventional wisdom or cliched advice. We must be willing to go wherever the data takes us,

---

* Seinfeld *fans may again be thinking of George Costanza. George was known for his signature breakup line, "It's not you, it's me." George considers this line so core to his romantic being that he flips out when a woman breaks up with him and uses the exact same line. (He finally wears her down until she admits that the breakup was, in fact, due to him. "All right, George. It's you," she tells him.)*

My point is that George's signature line can now be defended with data science. George could more robustly state his claim as follows: "According to machine learning models that predict romantic happiness, my mental state is four times more important in predicting my relationship happiness than everything about you. You know, scientists have found that it is extremely difficult to be happy in a relationship if you are not satisfied with life, suffer from depression, and have a negative affect— all traits that I have. Until I develop a more positive life outlook, it's going to be extremely difficult for me to be happy in a relationship with anybody. It's not you, it's me!"

even if that is to findings like those featured on Daily Inspirational Quotes.

So, as discovered by both a team of eighty-six scientists and whoever writes Daily Inspirational Quotes, one's own happiness outside a relationship is by far the biggest predictor of one's happiness in a romantic relationship. But what else predicts romantic happiness beyond one's own preexisting mental state? What qualities of a mate are predictive of romantic happiness? Let's start with the qualities of one's mate that are least predictive of romantic happiness.

LOOKS ARE OVERRATED—AND OTHER ADVICE THAT YOU HAVE LONG BEEN TOLD AND CONSISTENTLY IGNORED BUT MIGHT BE SLIGHTLY MORE LIKELY TO FOLLOW, KNOWING THAT DATA SCIENCE HAS CONFIRMED IT

Among more than 11,000 long-term couples, machine learning models found that the traits listed below, in a mate, were among the least predictive of happiness with that mate. Let's call these traits the Irrelevant Eight, as partners appear about as likely to end up happy in their relationship when they pair off with people with any combo of these traits:

» Race/ethnicity
» Religious affiliation
» Height
» Occupation

» Physical attractiveness
» Previous marital status
» Sexual tastes
» Similarity to oneself

What should we make of this list, the Irrelevant Eight? I was immediately struck by an overlap between the list of irrelevant traits and another data-driven list discussed in this chapter.

Recall that I had previously discussed the qualities that make people most desirable as romantic partners, according to Big Data from online dating sites. It turns out that that list—the qualities that are most valued in the dating market, according to Big Data from online dating sites—almost perfectly overlaps with the list of traits in a partner that don't correlate with long-term relationship happiness, according to the large dataset Joel and her coauthors analyzed.

Consider, say, conventional attractiveness. Beauty, you will recall, is the single most valued trait in the dating market; Hitsch, Hortaçsu, and Ariely found in their study of tens of thousands of single people on an online dating site that who receives messages and who has their messages responded to can, to a large degree, be explained by how conventionally attractive they are. But Joel and her coauthors found, in their study of more than 11,000 long-term couples, that the conventional attractiveness of one's partner *does not* predict romantic happiness. Similarly, tall men, men with sexy occupations, people of certain races, and people who remind others of themselves are valued tremendously in the dating market. (See: the evidence from earlier in this chapter.) But ask thousands of long-term couples and there is no evidence that people who succeeded in pairing off with mates with these desired traits are any happier in their relationship.

If I had to sum up, in one sentence, the most important finding in the field of relationship science, thanks to these Big Data studies, it would be something like as follows (call it the First Law of Love): *In the dating market, people compete ferociously for mates with qualities that do not increase one's chances of romantic happiness.*

Moreover, if I had to define the qualities that are highly desired even though they don't lead to long-term romantic happiness, I would call many of them shiny qualities. Such qualities immediately grab our attention. Just about all of us are quickly drawn to the conventionally beautiful, for example. But these attention-grabbing, shiny qualities, the data suggests, make no difference to our long-term romantic happiness. The data suggests that single people are predictably tricked by shininess.

YOUKILIS OF LOVE: A DATA-DRIVEN EMPHASIS ON UNDERVALUED ASSETS

After poring through all this research in relationship science, it occurred to me that the dating market today has a striking similarity to the baseball market in the 1990s.

Recall the Moneyball revolution, which was the motivation for this book. The Oakland A's and a few other teams realized, thanks to data analysis, that the market for baseball was all off. There was a disconnect between the cost of players in the open market (think: the salary you would have to pay them) and the value players brought to your team (think: how many wins they can add).

Players were frequently drafted and paid based not on the value they were likely to bring to the team but on other factors. The baseball market tended to overvalue shiny qualities in baseball players, such as being good-looking, and undervalued players who didn't immediately look like they should be baseball stars.

One such undervalued player was Kevin Youkilis. Youkilis has been described as "a fat third baseman who couldn't run, throw, or field." His college coach explained, "He was kind of a square-shaped body, a guy [who] in a uniform didn't look all that athletic. He wasn't a tall, prospect-y looking guy. He looked chubby in a uniform." Despite incredible college statistics, Youkilis's odd-for-a-baseball-player appearance caused him to fall to the eighth round of the draft.

But data analysts knew that, despite not looking the part of a great professional baseball player, Youkilis had all the tools that really mattered. The Boston Red Sox, motivated by such data analysis, took him in the eighth round, to the great frustration of the A's general manager, Billy Beane. Beane desperately wanted him but thought he would fall even further. The relatively short and chubby pick would eventually become a three-time All-Star and help lead his team to two World Series championships.

Data-driven teams in the 1990s, in other words, had success by focusing their attention on players, like Youkilis, who lacked the shiny traits that excited teams that didn't know the data. As Michael Lewis put it, "The human mind played tricks on itself when it relied exclusively on what it saw, and every trick it played was a financial opportunity for someone who saw through the illusion to the reality."

Similarly, the data has revealed striking inefficiencies in the dating market, with the mind playing tricks on single people. There is a disconnect between the cost of potential matches (think: how difficult it is to get dates with them) and the value of potential matches (think: the chances that they will make you happy in a long-term relationship).

So, could you approach dating with a similar mindset as that of Billy Beane? Could you focus more of your dating attention on targets whom the rest of the dating market ignores, even though they are just as likely to be great romantic partners?

The following groups of people, data has proven, all are dramatically less competed over in dating, even though the evidence suggests they are just as capable of making a partner happy.

Massively Undervalued Groups in the Dating Market

» Short men
» Extremely tall women
» Asian men
» African-American women
» Men who are students or work in less desirable fields, such as education, hospitality, science, construction, or transportation
» Conventionally less attractive men and women

By focusing more of your romantic attention on these groups of people, you will face less competition for an amazing mate. You are more likely to find a great partner whom others incorrectly ignore. You may find your Youkilis of Love!

OF COURSE, TELLING PEOPLE TO CARE LESS ABOUT SHINY QUALities, such as conventional attractiveness, that are overly valued in the dating market, while it may be sound, data-driven advice, seems like difficult advice to follow. There is a reason that shiny qualities are desired: shininess, almost by definition, grabs our attention. Return to Adam Duritz's point that we are all looking for "something beautiful." Is there any evidence-based way to allow yourself to stop being fooled by shininess in your search for romantic fulfillment?

One important, relevant, fascinating, and data-driven finding was uncovered by researchers at the University of Texas. In the beginning of a course, the professors asked all the heterosexual students in that course to rate the attractiveness of each of their opposite-sex classmates. Not surprisingly, there was a good deal of consensus. Most people picked the same classmates as the most attractive; these people were, by definition, conventionally attractive. Think Brad Pitt or Natalie Portman or the closest equivalents in the class.

At the end of the course, professors again asked the students to rate the attractiveness of each of their opposite-sex classmates. This is where the study got interesting. Now there was more disagreement in the attractiveness ratings. At the end of the class, people were far more likely to rate a person that other people didn't find so attractive as the most attractive.

What happened between the beginning and the end of the course that led so many people to change the rankings of their classmates' attractiveness? The students spent time with each other.

The man with the hunter eyes and chiseled chin may have seemed attractive at the beginning of the class. But he became less attractive to people who didn't enjoy their time talking to him. The woman with the hooked nose and the low cheekbones may have seemed unattractive at the beginning of the class. But she became more attractive to people who enjoyed their time talking to her.

The results of the study have profound implications for how we approach dating. Recall that we tend to seek out mates with conventional physical attractiveness and some other shiny traits—the types of people who would have scored very high on attractiveness ratings on the first day of a course—even though these people are highly competed-after and don't make for better mates. When we come across people lacking these shiny traits, we tend to not feel attraction toward them and not go on dates with them.

The research suggests that we might overrule our initial lack of attraction. Physical attraction, research shows, can grow over time if we like a person (or disappear over time if we don't like a person). The data suggests we should go on more dates with undervalued assets (those who might not have the qualities that so many people find so alluring) even if we don't initially find them attractive—and be patient, allowing a potential attraction to grow.

So much for the qualities that don't predict relationship happiness, such as conventional attractiveness. What are the qualities that do?

THE MOST LIKELY BEST MATE: SOMEONE SATISFIED WITH LIFE, SECURE WITH WHO THEY ARE, WHO CONSCIENTIOUSLY TRIES TO BETTER THEMSELVES

Joel and her coauthors found a few qualities of partners that did have some predictive power in how happy their romantic partners were. Their research suggests the following qualities are among the most predictive of a good mate:

» Satisfaction with life
» Secure attachment style (Be patient if you do not know what this—and a few other phrases—mean; I will explain them shortly.)
» Conscientiousness
» Growth mindset

What should we make of this list?

The first lesson may be that, to up your odds of romantic happiness, you should . . . read obscure psychology journals to learn what their terms mean. It turns out that the best predictors of how happy you are likely to be with a romantic partner are how they score on various quizzes that psychologists have come up with. This means, next time your partner suggests that you should turn off the ballgame and join her on the couch to take some new psychology quiz she discovered on the internet—instead of throwing a big fit about how you hate these stupid psychology quizzes and just want to be left alone to watch sports for one freakin' night and maybe it is best being single in life anyway—you should join her. Then you can find out if she has the qualities that make for a good long-term mate. Better yet, you might suggest taking these quizzes yourself.

So what do these quizzes determine?

Satisfaction with life is self-explanatory. People who are satisfied with life tend to make for better long-term relationship partners. Warning: silly Mick Jagger semi-joke coming. When Jagger gets onstage and sings "I Can't Get No Satisfaction," his voice, rhythm, and charisma may be hot, but listen to the words and you will notice a red flag for his capability to make a woman happy in a long-term relationship.

Attachment styles are explained in the excellent book *Attached* by Amir Levine and Rachel Heller. People with secure attachment style—this is the ideal trait in a partner—can trust people and are trustworthy, are comfortable expressing interest and affection, and have an easier time being intimate with others. An attachment style test can be taken here: https://www.attachmentproject.com/attachment-style-quiz/.

Conscientiousness is one of the Big Five personality traits, first proposed by Ernest Tupes and Raymond Christal in 1961. Conscientious people are disciplined, efficient, organized, reliable, and, according to Joel and her coauthors, better long-term partners. A conscientiousness quiz can be taken here: https://www.truity.com/test/how-conscientious -are-you.

Growth mindset is a trait developed by the psychologist Carol Dweck. People who have a growth mindset tend to believe they can improve their talents and abilities through hard work and persistence. Such people may work to become better romantic partners, which may be why they end up being just that. A growth mindset test can be taken here: https://www.idrlabs.com/growth-mindset-fixed-mindset /test.php.

The traits most predictive of romantic happiness are quite striking—and have profound implications for how we think about the romantic market. Recall the frequently depressing data from online dating sites on the ruthless superficiality that daters show. People really, really want sexy mates, even though there aren't that many sexy mates to go around.

It would certainly have been possible, in the data from actual relationships, that people who ended up with sexy mates ended up happier. Perhaps they really felt long-term joy from the wild sex or were satisfied that they could impress people at parties with the hotness of their partner. But the data from thousands of couples shows that this just isn't so. If anybody is likely to end up happier, it is daters who picked a mate with nice character traits.

And you can learn from the romantic successes and failures of more than ten thousand other couples. In looking for a mate, don't judge people based on the color of their skin, the symmetry of their faces, the height of their bodies, the sexiness of their profession, or whether they happen to share your initials. In the long term, data tells us, it really is the content of their character that matters most.

THE SEEMINGLY RANDOM, UNPREDICTABLE CONNECTION BETWEEN TWO PEOPLE

Why do some couples get happier over time? Why do some relationships that start great fall apart over time?

Samantha Joel and her coauthors also tried to answer these questions. The researchers took advantage of the fact that, for many relationships in the dataset, the romantic partners had been surveyed multiple times, sometimes years apart. Some romantic partners reported that, whereas they started unhappy in their relationship, they became increasingly happy. Others reported the reverse. What do couples that get better over time tend to have in common? What about those that get worse?

As part of their groundbreaking study, Joel and her coauthors used machine learning on the extensive data they had collected on thousands of couples to try to predict *changes* in romantic happiness. Note that this is a different question from the one in the last project we discussed, in which the researchers tried to predict whether members of romantic couples were happy at a *particular point in time.*

So, what could enormous datasets and machine learning tell us about the long-term trajectory of relationships? What do a couple's demographics, values, psychological traits, and preferences tell us about whether their relationship will get better or worse over time?

Nothing.

Joel and her coauthors' models had virtually no predictive power in predicting *changes* in romantic happiness. Happy couples are more likely to be happy in the future. Unhappy couples are more likely to be unhappy in the future. But there is nothing else about the two people that could improve predictions about future happiness.

The failure of the predictive models here has, I would argue, important implications for people's romantic decisions.

Certainly, many people make romantic decisions based on projected changes in happiness. How many times has one of your friends stayed in a relationship in which they weren't happy because they think, on paper, they should be happy—and eventually will be happy? "Sure, I'm miserable now," the friend might say. "But this relationship *should* work. It *has* to get better."

The results suggest that people are largely making a mistake when they expect their happiness in a relationship to change based on various qualities of them and their partner. The friend who stays in a relationship in which he is unhappy because he thinks he and his partner have so much in common and will eventually be happy is making a mistake.

The data suggests there is no better way to predict your future romantic happiness than your current romantic happiness. And if a partner does not make you happy now, you are wrong to assume that, because of the qualities of you and your partner, you will be happy in the future.

Or, to sum up all the data-driven advice on how to pick a mate: When you are single, focus more of your romantic search energy on people who lack the traits that are highly competed over. Focus more of your attention on people with strong psychological traits. Once you are in a relationship, pay attention exclusively to how happy you are with your partner—and do not worry about or gain false confidence from the similarities or differences between you and your partner. Don't think you have some ability to recognize a currently good relationship that will go south or a currently bad relationship that will go north. If the world's greatest contemporary scientists, using the most comprehensive dataset ever assembled, can't predict these types of changes, you can't, either.

UP NEXT

If you find a mate, you may have kids. And, if you have kids, you will undoubtedly wonder how you can be a better parent. There are new, important insights in what makes a great parent in enormous datasets, particularly the tax records of hundreds of millions of Americans.

CHAPTER 2

LOCATION. LOCATION. LOCATION. THE SECRET TO GREAT PARENTING.

Parenting is, in a word, challenging. A recent study calculated that, in the first year of a baby's life, parents face 1,750 difficult decisions. These include what to name the baby, whether to breastfeed the baby, how to sleep-train the baby, what pediatrician to get the baby, and whether to post pics of the baby on social media.

And that is only year one! Parenting doesn't become easier after that. In fact, parents have ranked the age of eight as the most difficult year to parent.

How can parents make these thousands of difficult decisions? Of course, parents can always turn to Google, where they will find plenty of supposed answers to just about any parenting question they have. But conventional parenting advice tends to be split between the obvious and the conflicting.

Examples of the obvious: KidsHealth.org urges parents to "Be a Good Role Model" and "Show That Your Love Is Unconditional." Examples of the conflicting: recently, the *New York Times* published an article that recommended parents "Try timeouts" to discipline their kids. In 2016, *PBS NewsHour* published a column online, "Why you should never use timeouts on your kids."

A frustrated mother, Ava Neyer, ranted after reading large numbers of books on parenting, particularly on baby sleep and development:

> *Swaddle the baby tightly, but not too tightly. Put them on their backs to sleep, but don't let them be on their backs too long or they will be developmentally delayed. Give them a pacifier to reduce SIDS. Be careful about pacifiers because they can cause nursing problems and stop your baby from sleeping soundly. If your baby sleeps too soundly, they'll die of SIDS.*

Ava Neyer, I would be lying if I said I could relate. (I'm not a parent; I'm merely an uncle. My decision-making process largely consists of asking my mom what gift I should get my nephew and her telling me "get him a truck" and me getting him a truck and then my nephew thanking me for the next four years for once having gotten him a truck.)

Regardless, I've scoured the parenting literature to understand what data can tell Ava and all the other parents out there. Is there anything parents can learn that is both not obvious and not conflicting? Can science offer advice for the thousands of difficult decisions that parents must make?

While there is not yet a convincing science-driven answer

to every one of the 1,750 first-year parenting decisions or the thousands of decisions beyond that, there are two extremely important, scientifically proven, and non-obvious lessons from science about parenting.

» Lesson number one: the overall effect of most of the decisions that parents make add up to less than most people expect; this suggests that parents fret too much about the vast majority of decisions that they must make.

» Lesson number two: there is one decision that a parent makes that is the most important—and many parents make the wrong decision here. If a parent makes a great, data-driven decision on this choice, that by itself would make any parent a far above-average one.

We will explore these lessons—and the evidence for them—in turn.

THE OVERALL EFFECTS OF PARENTS

Let's start with the most basic question about parenting: How much do parents matter? How much can "great" parents improve a kid's life, compared to average parents?

You could imagine three different worlds.

World 1 (Great Parents Can Turn a Potential Flight Attendant into a Dental Hygienist)

In this world, great parents can raise a kid who would otherwise have a middle-income job, making about $59,000 per year (say, a plumber or flight attendant), and turn them into a person with a job that pays slightly above average, making $75,000 per year (perhaps as a registered nurse or dental hygienist).

World 2 (Great Parents Can Turn a Potential Flight Attendant into an Engineer)

In this world, great parents can raise a kid who would otherwise have a middle-income job, making about $59,000 per year (say, a plumber or flight attendant), and bring them into the upper middle class, making $100,000 per year (perhaps as an engineer or a judge).

World 3 (Great Parents Can Turn a Potential Flight Attendant into a Brain Surgeon)

In this world, great parents can raise a kid who would otherwise have a middle-income job, making about $59,000 per year (say, a plumber or flight attendant), and make them rich, earning $200,000 per year (perhaps as a surgeon or psychiatrist).

Many people think that we live in World 2 or World 3— that skilled parents can help propel just about any kid a few rungs up on the socioeconomic ladder.

And it is not in doubt that certain parents have raised more than their fair share of noteworthy kids. Just consider Benjamin and Marsha Emanuel—and their three sons, Ari, Ezekiel, and Rahm.

» Ari is a high-powered Hollywood agent who was the basis of Ari Gold in the HBO show *Entourage*.
» Ezekiel is a vice provost at the University of Pennsylvania.
» Rahm was the White House chief of staff to Barack Obama and the mayor of Chicago.

In other words, Benjamin and Marsha reared sons who reached the highest echelons of business, academia, and politics.

Now, I know what some of my Jewish readers are thinking after learning of the life outcomes of Benjamin and Marsha Emanuel's offspring. Some of you Jews are thinking, "Yeah. That's all well and good. But did the Emanuels rear a doctor?"

An old Jewish joke (1/n of old Jewish jokes in this book) goes as follows:

"The first Jew ever was elected president. At the inauguration, his mother is sitting amongst all the dignitaries, as he is being sworn in. She tells everyone around her, in a loud voice, 'See that fellow up there being sworn in? His brother is a doctor.'"

No need to worry. Ezekiel, in addition to his academic career, is an oncologist.

There is even a supposed lesson in parenting from the success of the Emanuel brothers, as Ezekiel wrote about their upbringing in his book, *Brothers Emanuel*.*

---

* Q: *What does a family do after all the kids become enormously accomplished and wealthy?* A: *Publish a best-selling book explaining how it happened—and get even more glory and money.*

In *Brothers Emanuel*, we learn that every Sunday, while most families would watch Chicago Bears games, the Emanuel family would go on a cultural excursion, perhaps to the Art Institute of Chicago or a musical. When the boys yearned for lessons in karate or jujitsu, their mother insisted they instead take lessons in ballet. All three were taunted by other kids for it but now think the experience helped build discipline, character, and fearlessness.

The seeming lesson from the Emanuels: Encourage your kids to be cultured and different. Make your boy wear tights even if other boys mock him.

But, in truth, a single family, no matter how accomplished, cannot prove the validity of any given parenting strategy. And it is easy to find counterexamples to the Emanuel lesson. Take, for example, Dale Fernsby.* Fernsby recently responded to a mother on Quora, the community question-and-answer site, who wanted advice on whether she should sign her son up for ballet. Fernsby noted that his mother had enrolled him in many artsy activities as a kid, even though he hated them and was bullied for them. He says he learned from this experience that he was not allowed to have his own opinions or identity. He believes it caused him to have low self-esteem, an inability to express himself, and resentment toward his mother.

One challenge with learning how parents influence their kids is that one anecdote can never tell us all that much. Should we learn from the Emanuel story or the Fernsby story?

---

* *The name has been changed to protect the user's privacy, as he appears to have erased his answer since publishing it.*

Another challenge with learning about parental influence: correlation doesn't imply causation.

For much of the twentieth century, scholars searched for correlations in reasonably sized datasets between parenting strategies and child outcomes. They found many. Some of these correlations were summarized in the excellent book *The Nurture Assumption,* by Judith Rich Harris. For example, kids whose parents read a lot to them tend to have great educational accomplishments.

But how much of these correlations is causal? There is a major confounding issue: genetics. You see, parents don't just give their kids museum visits or ballet lessons or books. They also give them DNA. Return to the correlation between reading to kids and having well-educated kids. Are the kids drawn to education because of the books their parents read? Or are both parent and child drawn to books and knowledge because of their genetics? Is it nature or nurture?

Many stories have shown the power of genes in driving people's outcomes. Consider the evidence from identical twins who were raised apart. These people share the exact same genes but none of the same upbringing. Take, for example, the identical twins Jim Lewis and Jim Springer, who were raised separately from the age of four weeks. They reunited at the age of thirty-nine and found that they were each six feet tall and weighed 180 pounds; bit their nails and had tension headaches; owned a dog when they were kids and named him Toy; went on family vacations at the same beach in Florida; had worked part-time in law enforcement; and liked Miller Lite beer and Salem cigarettes.

There was, however, one notable difference between the

two Jims. They had given different middle names to their firstborn children. Jim Lewis named his firstborn James Alan, while Jim Springer named his James Allan.

Had Mr. Lewis and Mr. Springer never met each other, they might have assumed their parents played big roles in creating some of their tastes. But it appears those interests were, to a large degree, coded in their DNA.

Steve Jobs, who was adopted, had his own epiphany in the importance of genetics when he met, for the first time, his biological sister, Mona Simpson, at the age of twenty-seven. He was struck by how similar they were, including having both risen to the top of a creative field. (Simpson is an award-winning novelist.) As Jobs told the *New York Times*, "I used to be way over on the nurture side, but I've swung way over to the nature side."

Even the Emanuel brothers' story, which, on its surface, seems to show the power of great parenting, has a little-known kicker that suggests nurture may not have been the driver of the three boys' success. After having their three biological kids, Benjamin and Marsha Emanuel adopted a fourth child, Shoshana. Despite sharing the same cultural exposure of her three brothers, she did not share their genes—and has not had the same success.*

Is there any scientific way to determine just how much parents can affect their kids? To test the *causal* effects of parents on kids, we would seemingly have to randomly assign different kids to different parents—and study how they turned

---

* *In the next chapter, we will explore even more the relationship between genetics and success—and how you might use it to your advantage.*

out. It turns out, this has been done. The first convincing evidence on the overall effects of parents is all thanks to . . . a documentary on the Korean War.

One day, in 1954, an Oregon couple, Harry and Bertha Holt, parents to six children, saw a documentary on a theme they previously knew little about: Korean "G.I. Babies." These kids lost their parents during the Korean War and were now being raised in orphanages—with a shortage of food and love.

Now, Harry and Bertha Holt did not respond to this documentary the way that I tend to respond to documentaries: by zoning out during the entire film and then desperately trying to fool my girlfriend into thinking I had any idea what it was about. No, the Holts responded to the documentary on Korean orphans by deciding that they wanted to go to Korea and adopt many of these orphans.

The Holts' ambitious plan to adopt many orphans they had just seen in a documentary faced just one obstacle: the law. At the time, American laws allowed people to adopt at most two foreign children.

This obstacle proved fleeting. The Holts lobbied Congress to change the law. Congresspeople, impressed by the Holts' desire to do good, were persuaded. The Holts went to Korea. And, in a short time, they were back in Oregon—accompanied by eight new kids. The Holts were now a family of sixteen!

Soon news organizations covered the Holts' story. Radio stations told it. Newspapers wrote of it. Television stations broadcast it.

Further, just as the Holts were moved to action when they first learned the story of the "G.I. Babies," thousands of Americans were moved to action when they learned the story of the Holts. One after another, Americans said they wanted to follow in the path of the Holts. They wanted to adopt orphans as well.

Enter Holt International Children's Services, a foundation that makes it easier for Americans to adopt foreign babies. Over the years, more than thirty thousand Korean children have been adopted into the United States thanks to this organization. Parents merely sign up, get approved, and then get the next available child.

What does this story have to do with the science of parenting? Well, Bruce Sacerdote, a Dartmouth College economist, heard about the Holt program. Like so many other Americans, he was motivated to do something. In fact, he was motivated to run regressions!

You see, the process Holt uses to assign children to parents is essentially random, which means scientists have an easy way to test the effects of parents. They can simply compare adopted brothers and sisters who were randomly assigned to the same parents. The more the parents can influence the child, the more these brothers and sisters will end up alike. And, unlike with studies of genetically related children, we don't have to worry about genetics driving any correlations.

One cool thing about Sacerdote's study of the Holt program is that it allows us to see the effects of parenting, which we will get to in a bit. Another cool thing about the study is that we can see the difference between how nonprofit leaders and an economist describe the same foundation.

First, here is the description of Holt International Children's Services, from the organization itself. They report that they are "bringing light to the darkest situations" and "help strengthen vulnerable families, care for orphans, and find adoptive families for kids."

Next, here is the description of Holt International Children's Services by the economist, Sacerdote:

> *The random assignment of adoptees to families ensures that birth mother's education is uncorrelated with adoptive mother's education . . . [T]herefore $\beta 1$ is not biased by the omission of the first and third terms in (1).*

Holt International Children's Services believes they are "bringing light to the darkest situations." Sacerdote thinks they are making sure "$\beta 1$ is not biased by the omission of the first and third terms in (1)." I would argue that they are both right!

Anyway, what did that unbiased $\beta 1$ tell us? In most cases, that the family a kid is raised in has surprisingly little impact on how that kid ends up. Adopted brothers and sisters who were essentially randomly assigned to be raised in the same home end up only a little more similar than adopted brothers and sisters who were raised separately.

Remember, earlier I said there were three possible worlds, each representing a different degree to which parents might influence their kids. Sacerdote's study suggests that we live in World 1, the one in which parents don't have an enormous impact. A one standard deviation increase in the environment in which a child is raised, Sacerdote found, might raise a child's adult income by about 26 percent—not nothing but not too many rungs up the socioeconomic latter. Further, Sacerdote found the effects of nature on a child's income were some 2.5 times larger than the effects of nurture.

Sacerdote's study was just part of the evidence on the surprisingly limited effects of parenting. Other researchers have done further studies of adoptees. Researchers have also developed an ingenious method involving twins that allows them to disentangle the effects of nature and nurture, a method that I will explain in the next chapter.

Over and over, these studies converged on similar results, results that were summarized by Bryan Caplan in his provocative book, *Selfish Reasons to Have More Kids*: "Twin and adoption studies find that the long-run effects of parenting are shockingly small."

Parents, as surprising as it seems, and as the best evidence on the topic suggests, have only small effects on:

» Life expectancy
» Overall health
» Education
» Religiosity
» Adult income

They do have moderate effects on:

» Religious affiliation
» Drug and alcohol use and sexual behavior,
  particularly during the teens
» How kids feel about their parents

There are, of course, extreme examples in which parents can have an enormous impact on things like education and income. Consider the billionaire Charles Kushner, who gave $2.5 million to Harvard University—which likely caused the school to accept his son Jared despite a fairly low high school

GPA and SAT scores—and then gave Jared a stake in his lucrative real estate business. It is fair to say that Jared's educational achievements and wealth are far greater than they would be if he had had a different father. At the risk of being presumptuous, I think it is clear that Kushner's estimated $800 million net worth is many times higher than it would have been had he not inherited a real estate empire. But the data suggests that the average parent—the one deciding between, say, how much to read to their kids, rather than how many millions to give to Harvard—has limited effects on a kid's education and income.

If the overall effects of parenting are smaller than we expect, this suggests that the effects of individual parenting decisions are likely to be smaller than we expect. Think about it this way: if parents face thousands of decisions and the parents who make far better decisions only have kids who turn out some 26 percent more accomplished, each of the thousands of decisions, by itself, can't make a large difference.

Indeed, the best studies—many of them discussed in important books by Emily Oster—have generally failed to find much effect from even the most-debated parenting techniques. Some examples:

» The only randomized controlled trial on breastfeeding found that it had no significant long-term effect on a variety of child outcomes.
» A careful study of television use found that exposure to TV had no long-term effects on child test scores.
» A careful randomized trial suggests that teaching kids cognitively demanding games, such as chess, doesn't make them smarter in the long term.

» A careful meta-analysis of bilingual education finds that it only has small effects on various measures of a child's cognitive performance, and the effects may be entirely due to a bias in favor of publishing positive results.

Also, as it relates to the Emanuel/Fernsby debate about the merits of getting ballet lessons for boys, a meta-analysis found "limited evidence" that participation in dance programs may reduce anxiety; however, the authors suggest that this may be due to studies that are of "low methodological quality" and the results "should be treated with caution."

Look at careful studies rather than the latest attention-grabbing study, and you find that many of the things that parents worry about most turn out to have surprisingly little effects on their kids. Quite simply, most of the decisions that parents make matter less than they expect—and less than the parenting-industrial complex would like us to believe.

As Caplan put it,

> If your child had grown up in a very different family— or if you had been a very different parent—he probably would have turned out about the same. You don't have to live up to the exhausting standards of the Supermom and Superdad next door. Instead, you can raise your kids in the way that feels comfortable for you, and stop worrying. They'll still turn out fine.

Or as Caplan titled one of his sections of his book, offering his best advice to parents based on a couple of decades of social science research: "Lighten Up."

That, I would say, would be the best read of scientific-based parenting advice circa 2011, when Caplan wrote his book. Since 2011, the evidence has continued to accumulate that the overall effects of everything a parent does are smaller than most expect and that most of the decisions parents worry about don't have a measurable impact on how a kid turns out. However, there is an important update now, as there is some evidence that one decision that parents make may be the most important by far—and worth deeper consideration.

I would now advise parents: "Lighten Up . . . Except for One Choice You Make."

THE EFFECTS OF A NEIGHBORHOOD

"Asiyefunzwa na mamaye hufunzwa na ulimwengu."

Those are the words of my favorite African proverb. It translates, from Swahili, roughly as follows: "It takes a village to raise a child."

In case anyone is curious, my other favorite African proverbs (in English) are:

» "Rain does not fall on one roof alone."
» "Not everyone who chased the zebra caught it; but he who caught it, chased it."
» "No matter how hot your anger is, it cannot cook yams."

But back to "It takes a village to raise a child."

In January 1996, Hillary Clinton, then the first lady of the United States, expanded this proverb into a book, *It Takes a*

Village: And Other Lessons Children Teach Us. Clinton's book— and that African proverb—argue that a child's life is shaped by many people in the child's neighborhood: the firefighters and police officers, mail carriers and garbage collectors, teachers and coaches.

At first, Clinton's book seemed like another in a long line of entirely uncontroversial books that politicians write before seeking higher office, joining the likes of John F. Kennedy's 1956 profiles of courageous people, George H. W. Bush's 1987 recommendation to "look forward," and Jimmy Carter's 1975 defense of doing one's best.

However, a few months after Clinton's book was published, Bob Dole, the 1996 Republican nominee for president, thought he might be able to capitalize on the negative feelings many held towards the then first lady. And he eyed a seeming weakness in Clinton's seemingly incontestable thesis. Dole argued that Clinton's book, by emphasizing the important role that community members can play in a child's life, minimized the responsibility that parents have to raise these children. Clinton's argument, Dole claimed, was actually a subtle attack on family values. At the Republican convention, Dole pounced. "With all due respect," Dole said, "I am here to tell you: it does not take a village to raise a child. It takes a family to raise a child." The crowd roared. And that, friends, is the story of how the largest ovation at the 1996 Republican convention was devoted to an attack on a beautiful, moving, and touching African proverb.

So, who was right, Bob Dole or Africa?

For twenty-two years, the honest answer by data-minded scholars would be . . . (shoulder shrug). There hasn't been conclusive research one way or the other. The problem, once again: that difficulty establishing causality.

Sure, some neighborhoods produce more successful kids. Here's a fun fact that I talked about in my last book: Among Baby Boomers, one in every 864 kids born in Washtenaw, Michigan, the county that includes the University of Michigan, achieved something notable enough to warrant an entry in Wikipedia. One in 31,167 kids born in Harlan, Kentucky, a largely rural county, made it to Wikipedia. But how much of this is due to the kids of professors and other upper-middle-class professionals being really smart and ambitious—intelligence and drive they also would have used had they been born in rural Kentucky? Quite simply, the populations born in different neighborhoods are different, making it seemingly impossible to know how much a given neighborhood is *causing* its kids to succeed.

The shoulder-shrug-best-response to the effects of a neighborhood was the case until roughly five years ago. That was when the economist Raj Chetty began looking at this question.

RAJ CHETTY IS A GENIUS. DON'T BELIEVE ME? BELIEVE THE MacArthur Foundation, who in 2012 awarded him their "Genius Grant." Or believe the economics profession, who in 2013 gave him their John Bates Clark Medal for the best economist under the age of forty. Or believe the government of India, who in 2015 gave him the Padma Shri, one of their highest honors. Or believe the economist Tyler Cowen, who has called Chetty "the single most influential economist in the world today."

So, yeah, basically everybody is united in the belief that Chetty, who got his BA from Harvard in three years and his

PhD three years later and now ping-pongs between teaching at Stanford and Harvard, is extraordinary. (Chetty was a professor of mine in my PhD program at Harvard.)

A short while ago, Chetty and a team of researchers—including Nathaniel Hendren, Emmanuel Saez, and Patrick Kline—were given by the Internal Revenue Service de-identified and anonymized data on the complete universe of American taxpayers. Most important, by linking the tax records of children and their parents, Chetty and his team knew where people spent every year of their childhood—and how much they ended up earning as adults. If a kid spent the first five years of her life in Los Angeles and then the rest of her childhood in Denver, Chetty and his team knew that. They knew that not for a small sample of people; they knew it for the entire population of Americans. It was an extraordinary dataset in the hands of an extraordinary mind.

What can you do with this data on the entire universe of taxpayers to uncover neighborhood effects? Now, the naïve thing you could do is just bluntly compare the adult incomes of people who grew up in different places. But this would run into the problem already discussed: correlation does not imply causation.

This is where Chetty's cleverness—or, in the judgment of the MacArthur Foundation, genius—came into play. The team's trick was to focus on a particular, very interesting subset of Americans: siblings who moved as kids. Because the dataset was so large—remember, they were looking at every American taxpayer—they had a substantial number of such people to study.

How do siblings who moved as kids help establish the causal effects of neighborhoods? Let's think through how this might work.

Take a hypothetical family of two children, Sarah and Emily Johnson, and two cities, Los Angeles and Denver. Suppose, in our hypothetical world, when Sarah was thirteen years old and Emily was eight years old, the family moved from Los Angeles to Denver. Suppose further that Denver is a better place to raise a kid than Los Angeles. If this is the case, we would expect Emily to do better than Sarah because she had five more years in the Denver good-for-children air.

Now, of course, even if Denver were, on average, the better neighborhood to raise a kid in than Los Angeles, it is not 100 percent certain that Emily, with her five more years in Denver, would end up better off. Perhaps Sarah has some other advantage that overcomes the disadvantage of having spent less time in Denver. Perhaps Sarah was the smarter of the Johnson children, and her intelligence allowed her to outshine her sister.*

If you examined tens of thousands of these movers—as can be done using data from the entire universe of American taxpayers—the differences in the siblings would cancel out. In a sense, every time a family with at least two kids moves from one neighborhood to another, they are offering a test of the two neighborhoods. If the neighborhood they moved out of was the better neighborhood to raise a kid in, the older kid would be expected to do better—as that kid got more years in that neighborhood. If the neighborhood they moved into was the better neighborhood to raise a kid in, the younger kid would be expected to do better—as that kid got more years in that neighborhood. Again, this won't be true every time.

---

* *If you think that my saying that Sarah is the smarter of the Johnson children might be mean and unfair to Emily, please remember that these are made-up people.*

But, if you have enough movers, and certain neighborhoods are better to raise kids, you should see systematic differences in how younger versus older siblings perform when the family has moved into or out of that neighborhood.

Also, since siblings have the same parents and, in expectation, the same genetic capabilities, we can be confident that it is the neighborhood driving any consistent differences in how younger or older siblings do. Multiply over the entire universe of taxpayers, and add some clever math, and you have a measure of the value of every neighborhood in the United States.

So, what did the researchers find? Start with their analysis of large metropolitan areas. Consistently, some large metropolitan areas give kids an edge. If a child moves to the right neighborhood, they are less likely to end up in jail. They get better educations. They earn more money. Chetty and his coauthors found that being raised in the best cities (let's call them SuperMetros) can increase a child's income by about 12 percent.

Here are the five best metropolitan areas, those that offered, on average, the most improvement to a child.

SUPERMETROS

	Average increase in adult income from growing up here (compared to growing up in an average place)
Seattle, Washington	11.6%
Minneapolis, Minnesota	9.7%
Salt Lake City, Utah	9.2%
Reading, Pennsylvania	9.1%
Madison, Wisconsin	7.4%

Source: Equality of Opportunity Project

So, a parent might be wise to consider some of these metropolitan areas as great places to raise kids. However, parents
don't merely pick a metropolitan area to live in. They also pick
neighborhoods within these areas.

Chetty and his team's work has advanced beyond just telling us how good every metropolitan area is. Within a given
metropolitan area, some neighborhoods deliver better outcomes and others worse. And some neighborhoods may be
more advantageous to some groups than others.

In a remarkable paper, Chetty and coauthors used the tax
data to study how good every tiny neighborhood in the United
States was likely to be to raise a boy or a girl, kids of different
races, and kids of different socioeconomic status.

They found that there were large differences within metropolitan areas, with some neighborhoods giving dramatic
boosts to kids who grew up there.

For example, they broke down the data for Seattle and found
how kids raised in every census tract in the Seattle neighborhood
turned out. They found, for example, that North Queen Anne,
Seattle, was an advantageous place to raise a lower-income kid
whereas West Woodland was not. Overall, a one standard deviation improvement in neighborhood quality growing up might
increase someone's income by some 13 percent.

The researchers have created a website, http://opportunity
atlas.org, that allows anyone to find out how advantageous any
neighborhood is expected to be for kids of different income
levels, genders, and races.

The picture that follows shows the expected adult incomes
for a male of low-income parents being raised in different
neighborhoods in Seattle.

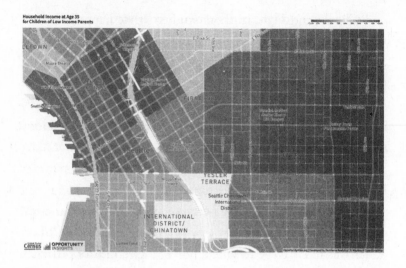

NEIGHBORHOODS MATTER MOST: THE MATH

Something interesting happens when we compare the work of Sacerdote and others (on the effects of parents) and the work of Chetty and his coauthors (on the effects of neighborhoods). The point is a bit subtle.

Recall that Sacerdote studied what happened when kids were essentially randomly adopted by different homes—and found that being raised in great homes would raise their income by about 26 percent. Now, a great home includes many factors: thousands of decisions parents made in how to raise them along with . . . the neighborhood that home was in.

Chetty and his coauthors studied what happened when kids moved to spend their formative years in different neighborhoods, without changing anything about the parents. They found that switching the same kid with the same parents to the best neighborhood could raise a kid's income by a significant fraction of the overall effect that Sacerdote found parents have.

If these studies are both correct (and they were very carefully done), this means that one factor about a home (the neighborhood it is in) accounts for a significant fraction of the effect of that home.

In fact, I have estimated, putting together the different numbers, that some 25 percent—and possibly more—of the overall effects of a parent are driven by where that parent raises their child. In other words, one of the thousands of difficult decisions a parent must make has much more impact than others. And for such an important decision, it hasn't featured much in parenting self-help books. In the book *The Parent Trap*, Nate Hilger has pointed out that, among sixty top parenting books, not one advises parents on where they should raise their kids.

If where you raise your kids is such an important decision, it would seem useful to have information of what qualities the places that are best for raising kids tend to share. Chetty and his coauthors studied that, as well.

WHAT MAKES GREAT NEIGHBORHOODS GREAT?

Once Chetty and his coauthors had a dataset of how every neighborhood ranked in how advantageous or disadvantageous it was to raise a child there, they could compare it to other datasets on neighborhoods to see what factors are most predictive of a neighborhood being a good one to raise a kid in. The neighborhoods that caused kids to have the best outcomes tended to consistently score high on a select few other variables.

If you are the sort of person who likes guessing games, see if you can figure out which aspects of a neighborhood predicted that kids who are raised there tend to do very well in life. On the list below of eight characteristics of neighborhoods, three were most predictive of giving kids a big boost in life, while five were not as highly predictive.

Which Three Characteristics of a Neighborhood Are Big Predictors That They Are Great Neighborhoods (People Raised There Are Most Likely to Do Well)?

» Number of High-Paying Jobs Nearby
» Percent of Residents Who Are College Graduates
» Strong Job Growth in the Area
» Student/Teacher Ratio of Schools in Area
» School Expenditures per Student in Area
» Percent of Two-Parent Households
» Percent of People Who Return Their Census Forms
» Population Density (Whether the Area Is Urban, Suburban, or Rural)

Got your guesses?

The three big predictors that a neighborhood will increase a child's success are:

» Percent of Residents Who Are College Graduates
» Percent of Two-Parent Households
» Percent of People Who Return Their Census Forms

Regardless of how well you did, can you now think what these three factors all have in common and what they tell us about what causes a neighborhood to be a good place to raise kids?

These three factors all relate to adults who live in that neighborhood. Adults who are college graduates tend to be smart and accomplished. Adults who live together in two-parent homes tend to have stable family lives. Adults who return their census forms tend to be active citizens.

This suggests that the adults that a kid is exposed to may have an outsized impact on how that child turns out. Of course, the correlation between qualities of the adults in the neighborhood and how the kids turn out doesn't by itself prove that the adults caused this outcome. But further research by Chetty and coauthors suggests that the adults a kid is exposed to in a neighborhood can exert a powerful advantage on those kids. In fact, the right adult role models appear to be more influential than the right schools or booming economies.

CASE STUDY # 1: THE POWER OF FEMALE INVENTOR ROLE MODELS

In one study, called "Who Becomes an Inventor in America? The Importance of Exposure to Innovation," Chetty and others combined a large number of datasets—including tax records, patent records, and test score data—to try to determine what predicts which children will end up making extraordinary scientific contributions.

Some of the results were not surprising. Early test scores were a big predictor of becoming a successful inventor. Kids who score well on math at an early age are more likely to end up having patents as adults.

In the depressing and not-all-that-surprising domain, the researchers found that a child's gender and socioeconomic status can influence the odds of becoming a successful inventor. Sadly, African-American and female kids are less likely to become inventors than white and male kids with the same childhood test scores.

But a factor that played a surprisingly large role in predicting becoming an inventor: the adults in a neighborhood a child grew up in. Kids who, when they were young, moved to neighborhoods with many adult inventors were more likely to become inventors when they grew up. The effect is highly concentrated in the industry in which these adult inventors worked. Move to a neighborhood with lots of people who invent medical devices, and a kid is more likely to grow up and invent medical devices.

Strikingly, the effects of living near adult inventors were

gender-specific. Chetty and his coauthors found that, for females, the chances of becoming an inventor rose if they grew
up near adult female inventors. There was no effect on a female's chance of becoming an inventor from growing up near
adult male inventors.

Little girls who see a lot of successful female inventors around
them try—and often succeed—to emulate these women. If
you want your daughter to become an inventor, one of the best
things you can do is get her in the vicinity, when she is young, of
women who have become successful inventors themselves.

CASE STUDY # 2: THE POWER OF BLACK MALE ROLE MODELS

A second study by Chetty and coauthors looked at the predictors of Black mobility in the United States. Sadly, in the
United States, African-American males are less socially mobile
than Caucasian males. The chart on the next page shows that,
if a white male and Black male have parents who earn the
same income, we can expect the Black male to end up earning
substantially less money.

While Black males have low social mobility in just about
every neighborhood in the United States, there are some
neighborhoods where they do better. Compare, for example,
Queens Village, a part of New York City, and West End, a
part of Cincinnati. In Queens Village, a Black male born with
parents in the 25th percentile of income can expect to end up
with income in the 55.4th percentile. In West End, a Black

male born with parents in the 25th percentile can expect to end up in the 31.6th percentile.

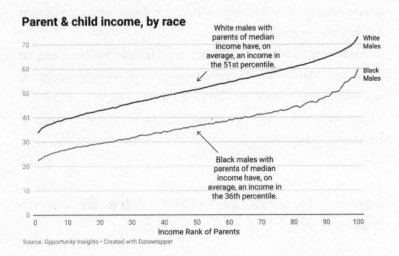

Parent & child income, by race

White males with parents of median income have, on average, an income in the 51st percentile.

White Males

Black Males

Black males with parents of median income have, on average, an income in the 36th percentile.

Income Rank of Parents

Source: Opportunity Insights • Created with Datawrapper

So what is it that explains the difference in the advancement of African-American males in neighborhoods?

There was a not-too-surprising and depressing variable that impacted African-American outcomes: racism. Chetty and his coauthors found that various measures of a neighborhood's racism, including racist searches on Google, negatively correlate with the advancement of African-American males. In *Everybody Lies*, I talked about my own research on the high levels of secret racism in the United States—as uncovered in Google searches. This is one more piece of evidence of the punishing nature of racism in the United States today.

But, once again, there was a more surprising factor that played an enormous role in how Black males turned out: adult role models. The researchers found that one of the biggest predictors of Black male achievement was how many Black fathers lived in the neighborhood. Return to Queens Village in New York City versus West End in Cincinnati. In Queens Village, 56.2 percent of Black boys are raised with a father around. In West End, 20.5 percent are.

And the importance of having Black male fathers around for a young Black male went well beyond the importance of having one's own father around. Even Black boys who grow up without their own father do far better when they are raised in places, like Queens Village, where many other Black fathers are around.

WHY ARE ADULT ROLE MODELS SO POWERFUL?

Why are the adults in a neighborhood so important in determining which girls pursue their dreams of becoming a scientist, which Black males escape the punishing reach of racism, and the life outcomes of so many others? How do we reconcile the surprisingly large effects of non-parental adults in a neighborhood with the sometimes surprisingly small effects of one's own parents?

One potential reason for this is that kids' feelings about their parents are complicated. Many kids rebel against their parents and aim to do the opposite of what their parents did.

If you are extremely well educated and a good citizen, perhaps your kids will be motivated to become the same themselves. But maybe they will be motivated to carve their own path and do the reverse.

But kids' relationships with the other adults in the neighborhood are much less complicated. There are no Oedipus or Electra complexes with the couple down the street. Kids are likely to see some of the other adults in town as people to admire—and emulate many of the things that they do.

Any parent will have a difficult time convincing their kids to act in the ways that they want them to act. But parents may find their kids naturally wanting to follow in the footsteps of some of the other adults that they see.

GIVING YOUR KIDS THE RIGHT ADULT ROLE MODELS

Some of the results in this chapter may be surprising. You may have thought that parents could have bigger overall impacts on their kids. You may have thought that living in Seattle wouldn't give a kid a significant income boost as compared to living in Los Angeles. You may have thought that the lady down the street would be unlikely to inspire your daughter's career.

And these results have important lessons for how to best parent your kids. The two major findings from the science of parenting, in fact, have different implications.

Finding one—the surprisingly small overall effects of parents, as uncovered by adoption and other studies—suggests that parents can really relax more on many of the decisions that they face.

If you are a parent who has a crisis many nights, unsure of what you should do, you are almost certainly approaching parenting with way too much anxiety.

In fact, for the large majority of parenting decisions, I might even condone a method of tackling them that you probably thought you would never hear from me: trust your gut. This is not because your gut has some magical powers that guide you to the correct answers. It is because the decisions just really aren't that big a deal. Doing what feels right and moving on is perfectly fine. The data, in some sense, justifies a simple, gut-driven approach. Be confident that, as long as you make some reasonable choices, you're doing about as well as parents can do.

But there is one area of parenting that, the data says, you may want to pay more attention to: the people you expose your kids to. You might be able to really influence your kids' lives here.

If you are really inspired by the work on neighborhoods, you can see how advantageous every neighborhood is on an interactive map found at the aforementioned https://www.opportunityatlas.org/.

But, even absent picking a neighborhood entirely based on data, you can use the spirit of the results in your parenting. Quite simply, it makes sense to expose your kids to adults whom you would want them to emulate. If there are people in your lives who you think might inspire your kids, engineer events in which your kids are exposed to them. Ask these possible role models to describe their lives and to offer advice to your kids.

There have been anecdotes suggesting that early role models for your kids can be powerful in shaping how they turn out. But data from tens of millions of Americans has proven it.

UP NEXT

Data in tax records can help you understand how to give a 12 percent or so boost to your kid's adult income. But what if you want to help your kids succeed in sports? Some new data can help you with that, as well.

CHAPTER 3

THE LIKELIEST PATH TO ATHLETIC GREATNESS IF YOU HAVE NO TALENT

What did you dream about becoming when you were a kid? For me, I had one dream and one dream only. I wanted to be a professional athlete.*

You see, I was obsessed with sports. And by obsessed, I mean *obsessed*. When I was four years old, my dad took me to a New York Knicks game, Julius Erving's final game at Madison Square Garden. The fans next to us, upon hearing me recite the players' stats, were convinced that I was a dwarf. It was impossible, they concluded, for someone so young to know so much about sports.

---

* *I also discussed my failed athletic dreams in* Everybody Lies, *where I wrote about the data on the surprising backgrounds of NBA players. Friends, this is a big theme in my life. And yes, it is somewhat therapeutic to study/write about it.*

But, if other people were impressed and charmed by my precocious sports knowledge, I was not happy with the situation I found myself in. My dream, as shared in paragraph one of this chapter, was to *be* a professional athlete, not to *know* the most about professional athletes. Why, I asked, couldn't I be Julius Erving and Julius Erving be the guy who knew my free throw percentage to the fourth decimal place?

But there was one seemingly insurmountable obstacle to my dream: I didn't have athletic talent. I was the shortest person in my class. Plus, I was noticeably slow. And, in case you are keeping score at home, yes, I was also weak.

Making my predicament that much more clear was the existence of my best friend, Garrett. Garrett was the tallest person in class, harboring a muscular physique and speed. Garrett was better than me at basketball, better than me at pitching, better at hitting, better at catching, better at soccer, better at running, better at dodgeball, better at Bonzo ball, and better at games I made up during recess to try to find something I was better at.

If my best friend from two blocks over was so much better than I at all these sports, how could I possibly become among the best in the world? I was a dreamer without a chance, a Seth who wished he were a Garrett, a jock trapped in a nerd's body. I was, in a word, screwed.

. . .

. . .

. . .

Or was I?

My father, Mitchell Stephens, an illustrious professor of journalism at New York University, felt genuinely sorry for me; it is hard to see your son want something so bad and have no way to achieve it. And he devoted his brainpower to trying to help me.

The Stephenses may not have height. We may not have speed. We may not have physical strength. But, goddammit, we do have cleverness! And, one afternoon, the cleverness rained down on Dad as he was watching a New York Jets game in his Old Navy pajama bottoms.

"The kicker!"

"That can't be that hard," Dad concluded. I could simply practice kicking footballs until I was world-class—and then my dream of being a professional athlete would be achieved.

The plan was in motion!

With proud smiles on our faces, the Stephenses went to Modell's Sporting Goods and got ourselves a kicking tee.

At first, I could barely get the ball off the ground. But I kept on practicing. And practicing. And practicing. Day and night. Snow, sleet, and rain. A boy, a tee, and a dream.

And that, friends, is the story of how I—a short and slow Jewish boy from the New Jersey suburbs—became a varsity kicker for the Stanford University football team!

Just kidding. That didn't happen, obviously. After months of practice, I was able to consistently kick the ball twenty-five feet in the air. Proud of my progress, one day I invited Garrett over to show him what I could do. Garrett said he wanted to see how far he could kick it. Despite having never kicked before, Garrett kicked it eighty feet in the air on his first try.

I quit trying to be an athlete. I began practicing math and writing, hoping maybe one day I could at least become world-class at analyzing what allows other men to achieve my dream.

Recently Discovered Photo of Data Scientist (Me) Trying (and Failing)
to Achieve His Childhood Dream

IN HIS BRILLIANT BOOK *THE SPORTS GENE,* THE SCIENCE JOUR-
nalist David Epstein started an extremely important conversation about what it takes to excel in sports. Epstein noted that,
while many parents and youngsters hope that passion and
hard work can lead to athletic greatness, increasing evidence
finds that genetics drive a large portion of athletic success.

This is perhaps clearest in basketball. I probably don't need
to tell you that height can be an advantage in basketball. You
may have noticed that. But you may not have noticed quite
how big an edge it can give people.

In fact, I and others have independently found that each
additional inch nearly doubles one's chances of reaching the
NBA. A 6'0" man has nearly twice the chances of reaching the

NBA as a 5'11" man. And the pattern holds throughout the height distribution. A 6'2" man has nearly twice the chance of reaching the NBA as a 6'1" man . . . A 6'10" man has nearly twice the chance of reaching the NBA as a 6'9" man . . . And so on.

These effects are so large that they imply that, while a man under 6 feet tall has about a 1 in 1.2 million chance of reaching the NBA, a man over 7 feet tall has a roughly 1 in 7 chance of reaching the NBA.

Further, Epstein points out that scientists have found that there are ideal body types—and genetic edges—in many other sports. The greatest athletes in the world tend to have been given, by the luck of the genetic lottery, bodies that are ideal for their sport. In swimming, for example, the ideal body tends to have a long torso and short legs, which allows for more torque with one's kicks. The best swimmers in the world tend to have genes that create these types of bodies. In middle- and long-distance running, by contrast, the ideal body type has long legs, which allows for longer strides. And the best of these runners tend to have genes that create these types of bodies.

Epstein has noted the striking contrast between the bodies of Michael Phelps, the most decorated swimmer of all time, and Hicham El Guerrouj, one of the greatest middle-distance runners of all time. Michael Phelps (at 6'4") stands seven inches taller than El Guerrouj (5'9"). But the legs of the two men are, incredibly, the same length. Or as Epstein has put it, "They wear the same length pants." Phelps's short legs helped him dominate swimming; El Guerrouj's long legs helped him dominate middle-distance running.

The findings that are presented in Epstein's book may be demoralizing for people, like me or others in my family, who dream of being world-class athletes but weren't given world-class athletic genes. Some parents or youngsters may read of such work and give up their dreams of athletic greatness. Why compete against the genetically gifted from all over the world?

But Epstein's book, as pathbreaking as it was, is merely the start of the conversation on what influences athletic success. Sure, genetics play an important role in athletic success.

But might it play a vastly different role in different sports? Are some sports almost entirely determined by your genes, whereas others are more likely to be determined by things like your passion and your hard work? Are there certain sports, as my dad hypothesized kicking in football to be, in which a young boy or girl without any great genetic advantage still has a chance of reaching the top through passion and hard work?

In a bit, I am going to show data that can give clues as to how much different sports rely on genetics—and which sports are best for someone who isn't genetically gifted. But before I get to those questions, I want to point out some awesome data uncovered by Patrick O'Rourke that is also relevant to the question of which sport might give a youngster without extraordinary skills the best shot of succeeding. O'Rourke wasn't necessarily looking for a sport that didn't require a genetic edge. He was looking for a sport that had many scholarship slots available per athlete.

FENCE YOUR WAY TO COLLEGE?

One evening, Patrick O'Rourke, a certified public accountant, was having dinner with friends and talking about his son, who was then a good high school baseball player but possibly not good enough to achieve a college scholarship. The friends came up with an idea for his son. Perhaps he should switch to playing lacrosse. After all, there are far fewer people who pick up the game of lacrosse. Perhaps he would have an easier shot of achieving a college scholarship if he focused his athletic energy on this less-played game.

O'Rourke was intrigued by this idea. But he did not merely accept what his friends told him. Instead, like a hero of this book, he began collecting data. For every sport, he collected data on how many high school players played the sport and how many scholarships were given out for that sport. He could then create an "Ease of Getting a Scholarship Athlete" metric: the percentage of high school athletes in a sport who get a scholarship.

So, what did the data say?

O'Rourke's friends were dead wrong. Sure, fewer men play high school lacrosse than high school baseball. But there are also far fewer college lacrosse scholarships than college baseball scholarships. Overall, the chance of a lacrosse player achieving a scholarship is 85:1. The chance of a baseball player achieving a scholarship is 60:1, a bit higher.

O'Rourke learned a lot more in the data, which he has shared with the world on his website, ScholarshipStats.com—and was first collated by the journalist Jason Notte.

ODDS OF GETTING COLLEGE SCHOLARSHIP FOR
MALES FOR DIFFERENT SPORTS

Sport	High School Athletes	College Scholarships Available	Ratio of High School Athletes to College Athletes
Gymnastics	1,995	101	20:1
Fencing	2,189	99	22:1
Ice Hockey	35,393	981	36:1
Football	1,122,024	25,918	43:1
Golf	152,647	2,998	51:1
Skiing—Alpine	5,593	107	52:1
Rifle	2,668	47	57:1
Basketball	541,054	9,504	57:1
Baseball	482,629	8,062	60:1
Soccer	417,419	6,152	68:1
Swimming & diving	138,373	1,994	69:1
Tennis	191,004	2,417	79:1
Lacrosse	106,720	1,251	85:1
Cross country	252,547	2,722	93:1
Track & field	653,971	5,930	110:1
Water polo	21,451	126	170:1
Wrestling	269,514	1,530	176:1
Volleyball	52,149	294	177:1

Source: ScholarshipStats.com; Table first created by Jason Notte for Marketplace.

ODDS OF GETTING COLLEGE SCHOLARSHIP FOR FEMALES FOR DIFFERENT SPORTS

Sport	High School Athletes	College Scholarships Available	Ratio of High School Athletes to College Athletes
Rowing	4,242	2,080	2:1
Equestrian	1,306	390	3:1
Rugby	322	36	9:1
Fencing	1,774	134	13:1
Ice hockey	9,150	612	15:1
Golf	72,172	3,056	24:1
Gymnastics	19,231	810	24:1
Skiing	4,541	133	34:1
Rifle	1,587	46	35:1
Soccer	374,564	9,266	40:1
Basketball	433,344	10,165	43:1
Lacrosse	81,969	1,779	46:1
Swimming & diving	165,779	3,550	47:1
Tennis	215,737	4,480	48:1
Softball	371,891	7,402	50:1
Volleyball	429,634	8,101	53:1
Field hockey	61,471	1,119	55:1
Water polo	18,899	344	55:1
Cross country	218,121	3,817	57:1
Track & field	545,011	8,536	64:1
Bowling	25,751	275	94:1

Source: ScholarshipStats.com; Table first created by Jason Notte for Marketplace.

These charts are striking. Who knew that a high school male gymnast had roughly nine times higher odds of getting a college scholarship than a high school male volleyball player? Or that a high school female rower had nearly a thirty-fold higher chance of getting a college scholarship than a high school female cross-country runner?

That said, O'Rourke has noted caveats with the data. Some of the sports that offer the best odds of scholarships, for example, have few high school programs and require joining pricey club teams. In addition, some scholarships are quite small. O'Rourke's website has much more information on every sport.

It would be wise, at least if you are American, to consult ScholarshipStats.com if you or your kid is considering specializing in a sport and dreams of playing that sport in college. However, it still will be tough to reach the top of many sports if you aren't genetically gifted, as Epstein famously told us.

So, what sports rely the most on genetics and what sports the least? I realized that the key to understanding how much genetics influence sports success in different fields could be found by calculating the . . . prevalence of identical twins in that sport.

TWINS: THEY'RE NOT JUST A BASEBALL TEAM IN MINNESOTA

Behavioral geneticists study why adults turn out the way they do—for example, why some people are Republicans and others Democrats. How much is nature and how much is nurture?

Disentangling these factors isn't so easy, however. The

central difficulty? The people who share nature also tend to share nurture.

Take siblings.

On average, siblings are more similar than random people in just about every dimension you could think to test. For example, siblings are far more likely than random people to share political beliefs. My younger brother, Noah, just about always agrees with my political analysis. We adore Barack Obama. We hate Donald Trump.

But why is this? Do Noah and I have coded in our DNA the same genes that make us moved by Obama's message of hope and change and turned off by Trump's message? This is certainly possible, since Noah and I share 50 percent of our DNA.

Or do Noah and I share political beliefs because our brains were both imprinted in similar ways at a young age? This is certainly possible, since many family dinners when we were kids focused on politics. And both our mom and dad supported Democrats. This pro-Democrat message was reinforced by our friends in the liberal neighborhood we were raised in outside New York City.

Noah and I share nature and nurture.

A German geneticist, Hermann Werner Siemens, came up with an ingenious solution to this problem. We can take an advantage of a natural experiment: twins.

About 4 times in every 1,000 pregnancies, one single egg is fertilized and splits into two separate embryos, yielding identical twins. These pairs of brothers or sisters will share 100 percent of their genes.

About 8 times in every 1,000 pregnancies, two separate eggs will be fertilized, at the same time, by two separate sperm, yielding fraternal twins. Same-sex fraternal twins, like identical twins, share the same birth date and, almost always, upbringing. But, unlike identical twins, fraternal twins only share, on average, 50 percent of their genes.

The nature/nurture debate can now be solved by some algebra equations, which I will spare you. The key point is this: if a trait is highly determined by genetics—in other words, if nature matters most—identical twins will be far more similar than fraternal twins. Of course, most traits are some combination of nature and nurture. But the exact contribution of each can be found from the equations.

In any case, these simple equations had large effects that rippled through society. For one, the realization of how valuable twins could be for behavioral research transformed the annual Twins Days Festival in the town of Twinsburg, Ohio.

Twinsburg was given its name in 1823. It was then that a pair of identical twin merchants, Moses and Aaron Wilcox, who owned land, money, and a sense of humor, made a deal with the town of Millsville, Ohio. The twins would donate to the town six acres of land, where a town square could be built, and twenty dollars, which could help build a new school. In return, the town would rename itself Twinsburg.

In 1976, the citizens of Twinsburg realized the perfect summer activity for a town with their name: a festival for twins. Twins from all over the world came. Some pairs of twins arrived with cool names, like Bernice and Vernice, Jeynaeha and Jeyvaeha, and Carolyn and Sharolyn. Some arrived with cool T-shirts such as, "Look out! There's 2 of me," "I'm the evil twin," or "I'm Eric, Not Deric." The twins held talent shows, a parade, and even a wedding. At the 1991 Twins Days Festival, Doug and Philip Malm, then thirty-four-year-old identical twins, met their wives, Jean and Jena, then twenty-four-year-old identical twins. They married two years later—at the 1993 Twins Days Festival.

Nothing could limit the fun of thousands of twins hanging out. Nothing, that is, except for scientists. Scientists got word that thousands of fraternal and identical twins would be in the same location on the same weekend. And they took off their lab coats, removed their goggles, got out their pencils and clipboards, and headed first thing to Twinsburg. The scientists transformed the annual Twins Days Festival from a weekend of fun and humor into a weekend of fun and humor—plus forms and tests.

Armed with their formulas for turning twin correlations into research papers, scientists offered a few bucks to the Twins Days Festival participants to help answer any question they could think to ask.

Want to know how much nature affects our trusting behavior? A team of scientists headed to the Twins Days Festival to find out. They had each twin play the Trust Game with another person, a game that sees whether players can cooperate to earn more money.

The scientists found that, compared to non-identical twins, identical twins are somewhat more likely to either both cooperate a lot, or both cooperate a little. In other words, identical twins tend to trust other people in similar ways. Plugging their data into the formulas, they found that variation in trusting behavior is 10 percent nature.

Want to know how much nature affects our ability to recognize sour tastes? A team of scientists headed to the Twins Days Festival to find out. They recruited 74 pairs of identical twins and 35 pairs of fraternal twins and had them drink various substances, of various concentrations of sour, and attempt to identify the taste. They compared the similarity of the point at which each pair of identical and fraternal twins could detect sourness. They plugged the numbers into the formulas and found that variation in the ability to recognize sour tastes is 53 percent nature.

Want to know how much nature affects people's tendency to be bullies? A team of scientists utilized mother and teacher reports of the bullying behavior of twins. They determined that bullying is explained 61 percent by nature.

Scientists have even located some genes that may be involved. For example, having the T allele at rs11126630 is consistently associated with a significant reduction in aggression during childhood—and, presumably, a reduction in the propensity to bully.

The thing I love most about that scientific finding is that it gives the ultimate comeback to a bully.

When I was a kid, a somewhat smart but extremely nasty bully told an effeminate nerdy boy, "You are missing a Y chromosome." The implication was that the nerd had a lack of manhood. The nerd could have come back with, "Well, you are missing a T allele at rs11126630." The implication would have been that the bully had an excess of aggression.

Over the past two decades, scientists around the world have used twins to calculate how much nature and nurture contribute to just about everything. Everything, that is, except developing world-class athletic ability in particular sports.

I decided to see if I could find data to help answer this question.

THE BASKETBALL GENES

If skill in a sport is highly dependent on genetics, science tells us that there will be a high prevalence of identical twins at the highest echelon of that sport.

Return to basketball. Basketball, in which success depends so much on the heavily genetic height, is a sport with a striking prevalence of identical twins playing at the highest level.

Since the advent of the National Basketball Association, ten pairs of twin brothers have made the NBA. At least nine of them have been identical.

In fact, assuming parents of NBA players birth identical twins at a roughly average rate, these numbers imply that an identical twin of an NBA player has more than a 50 percent chance of making the NBA himself. The average American male has a roughly 1 in 33,000 chance of making the NBA.

I built a model based on the twins equations that behavioral geneticists use to study other traits. (For the true geeks, mathematical details on the model, as well as the code, are on my website.) My best estimate is that variation in one's ability to play basketball is 75 percent nature. The ability to play basketball so well to make the NBA is really, really, really, really, really genetic.

Interestingly, it is possible that scouts do not quite realize just how all-important genetics are in basketball. A *Sports Illustrated* article discussed the difficulty scouts have evaluating identical twins. One Eastern Conference scout described scouting the (identical) Harrison twins, Aaron and Andrew: "They're very similar. I still don't even know which is which. The kid who hit the game-winning shots was the lesser of them. You have in your mind, one's the good one and one's the bad one. But then the bad one comes in and hits the game winners and you're like 'oh, shit!'" An executive suggested an interesting strategy for figuring out which prospect is better: "watch the mom." "Mothers," he noted, "always tend to cheer much more for the lesser player."

Three times scouts, perhaps by analyzing the responses of mothers, have been convinced that one identical twin prospect was clearly better than the other, ranking one at least twenty spots higher in the NBA draft. Each time, the twin judged inferior performed far more similar to his brother than their draft spots would have predicted.* Perhaps the scouts would have been better served ignoring the cheering patterns of the twins' mothers and merely assuming the twins would end up roughly the same. After all, they had the same DNA.

---

* *According to the formula by Aaron Barzilai found at https://www.82games.com/ barzilai1.htm, Jarron Collins (picked 52nd) would have been expected to have 16% as many Win Shares as Jason Collins (picked 18th). In fact, he had 78%. According to the same formula, Stephen Graham (undrafted) would have been expected to have fewer than 9.4% as many Win Shares as Joey Graham (drafted 16th). In fact, he had 21.8%. Caleb Martin (undrafted) would have been expected to have fewer than 27% as many Win Shares as Cody Martin (drafted 36th). In fact, he had 48% as many.*

While genetics are all-important in basketball—and trying to become a basketball player without the right genes is not a smart bet—they are less important in some other sports. Let's start with other major American sports.

THE (LESS IMPORTANT) BASEBALL AND FOOTBALL GENES

In baseball, there have been 19,969 people who have played in the major leagues—and some eight pairs of identical twins. This implies that an identical twin of a professional baseball player has roughly a 14 percent chance of making Major League Baseball, far smaller than the odds of an identical twin of a professional basketball player reaching the top. This is despite the fact that the odds of becoming a professional baseball player are three times higher than the odds of becoming a professional basketball player.

Football offers similar odds to identical twins as baseball does. The some twelve pairs of identical twins among 26,759 people who have played in the NFL implies about a 15 percent chance that an identical twin of a professional football player will become a professional football player himself.

The data is unambiguous that baseball or football skill relies less on genetics than basketball skill does. My best estimate is a genetic contribution to both baseball and football skill of around 25 percent.

In other words, genetics are less than half as important in football and baseball as they are in basketball.

THE ALMOST NONEXISTENT EQUESTRIAN AND DIVING GENES

We can extend the analysis to more sports—and again see that there are massive differences in just how important DNA is in different sports.

Bill Mallon, a former pro golfer and retired elbow surgeon, became obsessed with Olympic statistics. He is now a full-on Olympic historian and a stats provider for the International Olympic Committee. One of the stats that he collected: twins who have ever participated in the Olympics—and a rough estimate of whether they were identical. He was generous enough to provide these stats to me.

Some Olympic sports have shockingly large numbers of identical twins.

Take wrestling. Of 6,778 Olympic wrestling athletes, there have been something like thirteen pairs of identical twins. This implies that the identical twin of an Olympic wrestler has a better than 60 percent chance of becoming an Olympic wrestler himself.

Is this because identical twins wrestle with each other while growing up? Unlikely. The rates at which fraternal twins and same-sex siblings, who also could wrestle against each other, reach the Olympics are, I estimate, closer to 2 percent. Instead, the high prevalence of identical twins in wrestling suggests that genetics play an enormous role in wrestling talent. Other Olympic sports that have a high percentage of identical twins include rowing and track and field.

Some Olympic sports, however, Mallon's data shows, have

many fewer pairs of identical twins, suggesting genetics play a much smaller role in who reaches the top.

Take shooting. Of 7,424 Olympic athletes who have competed in shooting, there have been two pairs of identical twins. This implies that an identical twin of an Olympic shooter has a roughly 9 percent chance of becoming an Olympic shooter himself and suggests that shooting has only a small genetic component. And some sports—including diving, weight lift-

GENETICS OF SUCCESS CHART

	Percent of Same-Sex Siblings Who Are Identical Twins (Higher Number Suggests More Reliance on Genetics)
Olympic track-and-field athletes	22.4%
Olympic wrestlers	13.8%
Olympic rowers	12.4%
NBA players	11.5%
Olympic boxers	8.8%
Olympic gymnasts	8.1%
Olympic swimmers	6.5%
Olympic canoers	6.3%
Olympic fencers	4.5%
Olympic cyclists	5.1%
Olympic shooters	3.4%
NFL players	3.2%
MLB players	1.9%
Olympic Alpine skiers	1.7%
Olympic divers	0%
Olympic equestrian riders	0%
Olympic weightlifters	0%

Source: Author's calculations. Data for Olympic athletes was provided by Bill Mallon.

ing, and equestrianism—have had no pairs of identical twins. This suggests a small genetic component to skill in these sports. And it suggests that even someone who isn't given unusual genetic gifts might have a shot at rising in the sport with passion and hard work.

So what should we make of the sports that rely the least on genetics?

Certainly, some of these sports may be harder for the average person to break into. Equestrianism, for example, has famously high costs, which is why many children of rich kids specialize in the sport. Part of the reason equestrianism may have relied so little on genetics historically is that many people with unusual genetics gifts for riding horses—whatever those are—have never entered the sport.

That said, you can enter equestrianism without a lot of money these days and take advantage of the fact that passion and hard work are the predominant drivers of success in the sport. There are many websites, such as https://horserookie .com/how-ride-horses-on-budget/, that explain how to break into the sport without a lot of money.

AS I WAS STARING AT THE GENETICS OF SUCCESS CHART, I BEGAN thinking, as I often do, about Bruce Springsteen. This may be, in part, because I was listening to Bruce Springsteen, as I usually do while writing.

Springsteen's most famous song is "Born to Run." While the song is about the desire to escape a small town, the title also could describe the results from the identical twins analysis of sports. Track and field, after all, is among the sports most reliant on genetics.

Bruce Springsteen has a daughter Jessica. From the age of four Jessica Springsteen became obsessed with riding horses and she became one of the top equestrian athletes in the world, winning a silver medal at the Tokyo Olympic Games.

Bruce may be right that some of us are "Born to Run." But the data and the story of his daughter tell us that even those of us who weren't given freakish athletic skills can "Learn to Ride." And also learn to dive, lift, or shoot.

UP NEXT

Becoming a great athlete is one way to become rich. But there are other ways. New tax data is uncovering the secrets of who is wealthy in America—and they aren't always the people you expect.

CHAPTER 4

WHO IS SECRETLY RICH IN AMERICA?

Do you want to hear a boring story? (I sure know a catchy opening, don't I?)

Kevin Pierce* is a wholesale beer distributor. He is an owner of Beeraro, which was started in 1935, just after Prohibition, by his grandfather. In those days, wholesale beer distribution was an exciting industry. His grandfather has quipped that, when he started, whoever had the fastest car and the biggest gun sold the most beer. But these days, wholesale beer distribution is, like many businesses, a lot of spreadsheets and meetings.

At eight o'clock every morning, Kevin arrives at the four-hundred-square-foot office that he shares with a sales manager and team leader and starts looking at numbers: sales from the prior day and gross profit trends. Kevin might have a meeting or two with suppliers, where they haggle over prices. He might

---

* *Names and some details of this story have been changed.*

talk to a driver, particularly if that driver missed a delivery. He might talk with a team of pricing consultants he has hired—to help maximize the profits per drive.

While beer is consumed in the evening, the wholesale beer distribution business is a morning and afternoon endeavor. Most chain retail stores want their deliveries in the morning. Most bars and restaurants want their deliveries shortly before or after lunch. Thus, Kevin's workday always ends between 4 and 5 P.M.

Kevin is also making great money. Kevin says that he has made millions of dollars over the years from his business. It turns out Kevin just happens to be in one of the business fields in America with the highest shot of becoming a millionaire. A recent estimate from economists using newly available tax data found that wholesale beverage distribution is among a select group of industries in which a large percent of owners enter the top 0.1 percent of earners.

Kevin's money is consistent money, as well. In a good year, profits may be 2 or 3 percent higher than he had projected. In a bad year, profits may be 2 or 3 percent lower.

Kevin admits that his job can be "insanely boring" and he is "coming to hate spreadsheets." While beer does add at least a bit of sex appeal to his industry, he notes that his day-to-day job wouldn't be all that much different if he were selling toilet paper.

But Kevin is happy when he compares his career to those of his friends. Making great money year after year and ending work at five is a pretty good situation, he has come to realize. One of Kevin's friends recently commented on Kevin's life—the nice house, the control of his calendar, the consistent income. "I need a business like yours," the friend told Kevin.

Here's how Kevin sums up his career: "It's really boring. But every single day, we can make more and more money."

THE DATA OF RICH PEOPLE IN AMERICA

Who is rich in America?

The truth is, up until a few years ago, we had only limited answers to this question. Of course, we all had some idea that certain endeavors were more likely to lead to wealth than others. We had some sense that, say, people working at Goldman Sachs were making more money than teachers (whether or not Goldman Sachs employees *should* be making more money than teachers).

But, up until a few years ago, there had been no comprehensive study of the finances of *all* Americans, including an analysis of every rich person in the United States. Our understanding of who is rich in America has relied on two flawed approaches.

First, we can ask people. But this is complicated by the fact that many people do not want others to know how much money they have. Jack MacDonald, an attorney and investor in Washington, lived in a one-bedroom apartment, wore tattered shirts, and collected coupons to get better deals on groceries. When he died, he shocked everybody by donating $187.6 million that he had amassed from investments. On the flip side, Anna Sorokin moved to New York City in 2013 and immediately entered elite New York City social circles, telling people she was an heiress with a €60 million trust fund. She would eat at the finest restaurants and stay at the finest hotels, having friends pay and telling them she would pay them back. She was eventually arrested and imprisoned for scamming various people and institutions out of money, and it was revealed she was secretly broke. Most people aren't as extreme as MacDonald or Sorokin, but many people play up or down their wealth in the complex social dance of modern life.

The second way we tend to learn about rich people is we hear the stories of rich people, perhaps in the media. But the rich people we learn about are heavily biased toward the rich people whose lives make a good enough story to be told. Thus, our idea of rich people is massively slanted toward rich people who have a sexy story.

What happened a few years ago that allowed us to finally learn the full picture on who is rich in America? Academics have been allowed to work with the IRS to study digitized data on the entire universe of taxpayers in the United States. (All the data was de-identified and anonymous.) One team of researchers—Matthew Smith, Danny Yagan, Owen Zidar, and Eric Zwick, whom I will henceforth refer to as the Tax Data Researchers—have used this data to study the career paths of the entire universe of the most financially successful Americans.

Before discussing what they found, I must offer an important public service announcement, for fear of being yelled at by the ghosts of my now-deceased socialist Jewish grandparents: making huge sums of money is not necessarily the proper goal for life, nor does it necessarily make people happy. Chapters 8 and 9 will discuss what data science can teach us about happiness—and the (limited) role of money, among other factors, in its attainment.

RICH PEOPLE OWN

So, how do rich people make their money?

Let's begin with the not-that-incredibly-shocking: The majority of wealthy Americans, the Tax Data Researchers found, own a business; they don't make their money from a salary. More precisely, only about 20 percent of members of the top 0.1 percent receive the majority of their money from wages.* Among these rich Americans, 84 percent receive at least some money from owning their own business.

There are, for sure, some high-profile examples of rich people who attain their wealth by being paid a salary by an organization. Think of corporate CEOs, such as Jamie Dimon, who is paid more than $30 million per year by JPMorgan Chase; prominent network hosts, such as Lester Holt, who is paid more than $10 million per year by NBC; or coaches of top sports teams, such as David Shaw, who was paid $8.9 million in 2019 by Stanford University, despite once deciding—I kid you not—to punt from the opponent's 29-yard line.†

But these salary-based paths to wealth are, the data tells us, rare. The Tax Data Researchers discovered that, among the top 0.1 percent, for every 1 employee being paid a salary that

---

* *In the rest of the top 1 percent, about 40 percent receive the majority of their income from wages. To make the top 1 percent during the time period studied, someone had to earn $390K. To make the top 0.1%, someone had to earn $1.58M.*
† *This was an unnecessary rant that I expect nobody to appreciate except Gabe Rosen.*

puts them there, there are 3 owners collecting business profits that put them there. Or perhaps another way to think of this: among the wealthy, for every 1 Jamie Dimon (an employee) there are 3 Kevin Pierces (an owner).*

Even someone who does make a ridiculously high salary, such as a sports star, won't become as wealthy as someone who owns the right asset. Just consider this fun fact recently pointed out by the data scientist Nick Maggiulli. Of the more than 26,000 men who have played in the National Football League, do you know who has achieved the greatest wealth? That would be . . . Jerry Richardson. Who? The former wide receiver caught a grand total of fifteen passes in his two seasons in the NFL. But, shortly after retiring from football, he began a business career buying and expanding franchises of Hardee's, the fast-food restaurant. His stake in more than five hundred Hardee's outlets helped him build a net worth of more than $2 billion. In contrast, Jerry Rice, likely the greatest wide receiver of all time, earned an estimated $42.4 million in his twenty-year career in which he caught 1,549 passes. The math of wealth, in other words: 15 career receptions + 500 Hardee's is worth some 50 times more than 1,549 career receptions + 0 Hardee's.

By the way, Richardson used his wealth to become a founding owner of the Carolina Panthers but was forced to sell his stake in the franchise after allegations of using sexually suggestive language and a racial slur in the workplace. So,

---

* In the rest of the top 1 percent, a roughly equal number of people receive the majority of their income from wages as from owning a business.

make sure to take the right lesson from Richardson's story: the value of owning assets for building wealth, not how to be of good character.

RICH PEOPLE OWN THE RIGHT THING

While owning a business is the predominant path to becoming rich in the United States, it is no guarantee that you will become rich. For every Kevin Pierce, who consistently clears great money every year, there are many people who open a business that goes belly-up and many people who open a business that merely ekes out a small profit.

What separates the winners and losers in business? The field one enters is extremely important. Some fields make good businesses, with a good chance that an owner will become rich. Some fields do not.

Let's start with the fields in the latter category: those that don't make good businesses.

Tian Luo and Philip B. Stark examined which business fields give an owner the lowest chance of staying in business. They utilized an enormous dataset from the U.S. Bureau of Labor Statistics, which keeps track of every business operating in the United States.

The field with the single shortest average life span? Record stores. The average record store lasts 2.5 years before going bust. The typical record store, in other words, has a similar life path as many of the rock-and-roll giants that inspire the stores' founders: brief—minus, however, the groupies.

THE SEXY PATH TO QUICK FAILURE

Field	Median Time a Business in the Field Lasts (For Comparison, the Average Dentist's Business Lasts 19.5 Years)
Record stores	2.5 years
Amusement arcades	3.0
Hobby, toy, and game stores	3.25
Bookstores	3.75
Clothing stores	3.75
Cosmetics and beauty supply stores	4.0

Source: Luo and Stark (2014)

This was the start of a pattern uncovered in data: sexy businesses, the businesses that children might dream of starting, tend to fold fast. Also high on the quick-to-fold business list were amusement arcades, toy stores, bookstores, clothing stores, and beauty supply stores.

The study, in other words, offers an important caution for an aspiring entrepreneur. Be careful entering a field that is sexy or the stuff of childhood dreams.

In many sexy fields, the competition will be ferocious, and you will have no way to stand out from the pack. You risk losing a lot of money in a short amount of time. (There is, I should acknowledge, a more promising sexy path to business success, which I will discuss in a bit, but it requires a whole bunch of talent and hustle.)

SO, WHAT ARE THE FIELDS THAT ARE MOST LIKELY TO PRODUCE wealth?

The Tax Data Researchers documented how many businesses in every field put their owners in the top 0.1 percent of

TOP 5 BUSINESSES WITH THE GREATEST NUMBER OF MILLIONAIRES
(NOTE: THIS IS A MISLEADING CHART FOR PICKING A GREAT BUSINESS)

Field	Owners in the Top 0.1 %
Lessors of real estate	12,573
Activities related to real estate	10,911
Automobile dealers	5,236
Offices of physicians	4,711
Restaurants	4,471

Source: online appendix of Smith et al. (2019); this only includes data for S
Corporations.

earners in the United States. The chart is in Table J.3 of their
online appendix, for those curious.*

Above are the top five fields that create the most rich owners, although this chart is misleading for determining the best
paths to wealth for reasons I will discuss.

So, should everybody quit their jobs and start a business in
one of these fields that the Tax Data Researchers found have
the most rich owners? Is the path to wealth to start a restaurant and try to join the 4,000+ rich owners in that field? Is
someone who collects her grandma's recipes and opens a piz-
zeria not merely following her dreams but following the data?

No. The chart is misleading when it comes to understanding
good fields to start a business in, for the following reason: It shows
only the number of people who *got rich* starting a business in that
field. It doesn't consider the total number of people who *started a*

---

* *Boring, technical point: For the rest of this chapter, I am going to present data*
  on S-Corporations, the most popular corporate formulation in the United States.
  Unless otherwise noted, I am only talking about this business type.

business in that field. Some fields rank high merely because a huge number of people start businesses in that field, not because a large percentage of people who start businesses in that field get rich.

Return to restaurants. According to publicly available data from the census that I analyzed, restaurants are the single most popular business to start in the United States. There are more than 210,000 restaurants in the United States, meaning that the 4,471 rich restaurateurs make up some 2 percent of restaurant owners. In other words, starting a restaurant is not a particularly good bet to enter the ranks of the rich. If you have come across a lot of rich restaurateurs, this is in large part because there are so many restaurateurs.

Some fields, it turns out, give far better odds of wealth than that. Consider automobile dealerships, another field near the top of the Tax Data Researchers' chart. According to census data, there are only 25,200 automobile dealerships. This means that the 5,236 auto dealerships that minted rich owners represent a remarkable 20.1 percent of auto dealerships. In other words, the chances that a business put its owner in the top 0.1% was some ten times higher if it was an auto dealership than a restaurant.

COMBINING DATA FROM THE TAX DATA RESEARCHERS WITH PUB-licly available data from the census, I attempted to locate the most promising fields for becoming rich.* I located all business fields that fit two criteria:

---

* *For the very few of you interested in the details, I compared data from Appendix J.3 of the online appendix from the Tax Data Researchers, which records the number of S corporations with owners in the top 0.1 percent, broken down by industry, and SUSB Annual Data Tables by Establishment Industry from the United States Census Bureau, which records the total number of S corporations, broken down by industry.*

» First, *at least 1,500 business owners in this field are in the top 0.1 percent*. This means that many people have achieved wealth through this path.

» Second, *at least 10 percent of businesses in that field had an owner in the top 0.1 percent*. This ensures that many of the people who tried this path got wealthy.

Among hundreds of fields that people start businesses in, there were only seven fields that fit both criteria, having both high numbers of rich people as well as a high probability of wealth. Here they are:

THE GET-RICH CHART

Field	Percent of establishments with owner in the top 0.1%
Lessors of real estate	43.2%
Activities related to real estate	25.2%
Automobile dealers	20.8%
Other financial investment activities	18.5%
Independent artists, writers, and performers	12.5%
Other professional, scientific, and technical services	10.6%
Miscellaneous durable goods merchant wholesalers	10.0%

Sources: author's calculations based on data from online appendix of Smith et al. (2019) and the U.S. Census Bureau. These calculations are for S-Corporations.

Okay, so what should we make of this chart?

First, let's clarify what some of these fields are. Further research and conversations suggest that "other professional, scientific, and technical services" companies are largely market research firms; "miscellaneous durable goods merchant wholesalers" are middlemen, who buy goods in large quantities from manufacturers and then sell them to retailers.* The merchant wholesale business, the data suggests, is a great business. Wholesale beverage distributors, as already mentioned, are disproportionately members of the top 0.1 percent, although there aren't quite enough of them to make the Get-Rich Chart. And other goods merchants have better shots than most of getting rich, as well.

"Lessors of real estate" largely own real estate and rent it out; "activities related to real estate" are largely companies that manage real estate for others or appraise real estate. People in "other financial investment activities" largely manage and invest other people's money.

Combining the two fields of real estate and substituting out the technical names, we might say there is a Big Six of businesses that disproportionately make people rich.

The Big Six

» Real estate
» Investing
» Auto dealerships
» Independent creatives
» Market research
» Middlemen

What should we make of the Big Six?
Investing and real estate clearly aren't surprising mem-

---

* *This website gives information on how to start a wholesale distribution business: https://emergeapp.net/sales/how-to-start-a-distribution-business/.*

bers of the Big Six. But the other four fields among the Big Six were, at least to me, revelations. Middlemen? Auto dealerships? Market research? Certainly, nobody had ever told me that these fields were consistent paths to millionaire status.

To be honest, I hadn't even heard of middlemen; I associated auto dealers with shady hucksters; and I wouldn't have had any reason to believe that market research was far more lucrative than consulting (it is).

HOW CRAZY IS IT TO TRY TO BE A CELEBRITY?

I was shocked to see "independent artists, writers, and performers" on the list. We often think of artists as primarily broke. (Think of the cliche of the starving artist.) Also, the data tells us that most sexy businesses, such as record stores, fold quickly. So, what should we make of the fact that 12.5 percent of companies that consist of "independent artists, writers, and performers" have rich owners?

This is largely due to something called selection bias, an important bias that must be considered anytime that you analyze data. Most independent artists, writers, and performers don't have enough success that they turn themselves into a business for tax purposes. These struggling creatives, who never make any real profit, are not included in the data. If you added all these creatives, the odds of reaching big success as a creative would go way down. (This bias is much less of an issue for other businesses, in which a much higher percentage of people who start their own enterprise in the field will incorporate as a business.)

That said, even if selection bias artificially jacks up the odds of success of independent creatives, the true odds of having huge success in this field may have been higher than I had guessed. Sure, becoming rich as a creative is a long shot. But I might have guessed it was a 1-in-100,000 long shot. The odds may be less long than that.

Data from the Tax Data Researchers suggests there are at least 10,000 independent creatives who earn enough to be in the top 1 percent of earners. How many people try to be independent creatives?

There is conflicting information here. According to the Bureau of Labor Statistics, there are 51,880 "independent artists, writers, and performers," not all of whom own their own business. However, there are also other people who have a side hustle in a creative field but don't consider that their main profession. Other survey data suggests that 1.2 million Americans primarily make their living as working artists. This suggests the 10,000 independent creatives in the top 1 percent would make up roughly 1 percent of working artists. If this were true, the odds of entering the top 1 percent would be about as good for an artist as they would be for an average American. That said, there are some 2 million art graduates in the United States. To the extent all of them tried to make it as independent creatives and some of them have given up, perhaps 1 in 200 people who attempt a creative career make it big.

Clearly, more research needs to be done. But some of the new tax data combined with other data might suggest that art gives someone on the order of a 1-in-200 to a 1-in-100 shot at huge success. This might be too much of a long shot on its own to make it worthwhile for many people. But Chapter 6 shows that there are things artists can do to dramatically increase their chances of success. For example, merely presenting your art widely has been found to increase your chances of success 6-fold. If you follow the data-driven lessons on how to increase your odds of making it as an artist, you may get your odds up substantially, perhaps to 1 in 10. You still probably won't make it big, but that might not be a terrible bet for getting rich doing something you love.

Let's put it this way: before I saw this data, I would have said to anybody who dreamed of making it as an independent creative that that was a silly idea unless you had a trust fund to support you.

But, after seeing this data, I would be more equivocal. I would say trying to become an artist is a silly idea if you don't do the things that I discuss in Chapter 6 that maximize your chances of getting your lucky break. But if you do these things, trying to be an artist may not be a crazy bet, particularly when you are young. Another way to say this: trying to make it big as an independent creative but not producing a ton of work and not hustling to find your break is just about never going to work. If that's your plan, you're better off getting a degree in accounting. However, trying to make it big as an independent creative and producing a ton of work and hustling like crazy to find your break may give you a surprisingly decent shot of success. If that's your plan, you've got a shot, although you must be okay with the fact that you have a better-than-even chance of failing.

The data suggests that the chances of getting rich as an independent creative are higher than for some other businesses. This is particularly striking given the earlier finding that many sexy businesses, such as record stores, are among the least successful businesses. What does being an independent creative offer that starting a record store doesn't that may lead to a higher chance of making big profits?

Creatives join with the other members of the Big Six in pointing out what really makes a great field to start a business in: *great business fields allow for the existence of many local monopolies.*

THE ESCAPE FROM PRICE COMPETITION

Most businesses do not create many millionaire business owners.

For example, there are more than 49,000 gas station businesses in the United States; more than 15,000 dry-cleaning and laundry services businesses; and more than 8,000 death care services businesses. These are all nonsexy businesses that provide a fundamental service. But data tells us that they are not paths to being rich. Virtually none of the owners of these businesses are among the top earners in the United States.

In fact, the Tax Data Researchers data tells us that there are many fields in which tens of thousands of people start businesses, but few enter the top 1 percent, let alone the top 0.1 percent of earners.

Some Just-About-Never-Get-Rich Businesses

» Building equipment contractors
» Residential building construction
» Automotive repair and maintenance
» Services to buildings and dwellings
» Architectural, engineering, and related services
» Building finishing contractors
» Personal care services
» Gasoline stations

Why do the Big Six stand out among the other fields in creating such a high probability of their owners becoming rich?

It is time for a lesson in a field that is even more boring than many of the fields that mint many millionaires. It is time for a lesson from the subject of my PhD: economics.

Economics 101 tells us that a business's profit is, by definition, its revenue minus its costs. If a business can charge far more for its product than it costs to produce it, it will make a substantial profit. But, if a business cannot charge more for its product than it costs to produce it, it will make zero profit.

And Economics 101 tells us that it is surprisingly difficult for a business to make a profit. In fact, economics tells us that most businesses in most industries can expect to make little-to-no profit, which is why it is not at all surprising to economists that the Tax Data Researchers found so many industries in which so few business owners were getting rich.

Why is it so hard to make a profit?

Suppose there is a person, Sarah, who is running a business

that makes a large profit. Suppose it costs her $100 to make each unit of her product; every year, she can sell 10,000 units of this product for $200 each, netting her $1,000,000 per year.

Not bad, Sarah!

But Sarah would have a potential problem.

Suppose there is another person, Lara, who works at a dead-end job making $50,000 per year. She would love to have Sarah's profits. Suppose she quit her job and made Sarah's product for $100 but charged $150 instead. Now, all of Sarah's customers switch to Lara, due to the cheaper price. And Lara is making $500,000 per year. Sarah's new salary? Bupkis.

Lara is now raking it in, until Clara decides to join this business and try to get in on the profit action. She charges $125 for the product, steals all the customers, and makes $250,000 per year. Now Sarah and Lara are screwed.

What happens next? In this fictional example, the process of new people entering would abruptly end due to the fact that I ran out of rhyming names. In the real world, however, the process would continue, and Clara and every other business-person would have no protection from lack of rhymes.

This process would end only when nobody makes enough of a profit to incentivize any new people to enter the industry— or any of the firms already in the industry to lower their price. In Economics 101, this is called the *zero-profit condition*. Price competition will continue until the profits are driven down to zero.

Nobody thinking of starting a business should ever un-derestimate the power of the zero-profit condition. Many,

many people start businesses hoping to get rich only to find themselves stuck in ruthless price competition and eking out a living.

Consider the taxi driver who picked me up at a train station in upstate New York a few days ago and told me the story of his business career. Some twenty-five years ago, he got in the taxi business in the town he grew up in. He would wait by the train station and drive people who were coming back from work in New York City. For a while, this was a solid business, and he was making a solid living. But many competitors entered, including many poorer people who were willing to charge far less than he was. Now, when the train from New York City arrives, riders are accosted by a large number of cab-drivers. Customers generally just pick whoever will offer the lowest price.

The taxi driver's profits have largely disappeared. Moreover, when the Covid-19 pandemic hit, his business collapsed, forcing him to move back in with his parents.

To consistently keep profits, a business owner must somehow avoid competitors undercutting them on prices until the profits disappear. How can this be done?

Well, businesses in the Big Six all have a way to do it. Every field that creates many millionaires gives some way for their owners to avoid ruthless price competition that drives profits to zero.

Perhaps the most straightforward way to be insulated from price competition is by law. Law, in fact, is the explanation for the presence on the Big Six of the business that initially made very little sense to me: auto dealerships.

Auto dealerships are a heavily regulated industry. In many states, cars cannot be distributed by the companies that make them, although Tesla is currently challenging this rule. There are also limits that prevent new firms from dealing a particular company's cars.

Whether this is good for customers is debatable. But I'm not writing a legal treatise; I am writing a self-help book. And, clearly, if your goal is to be rich, having legal protection against competitors is a huge assistance. That is why owners of auto dealerships are so prevalent among rich people.

Recall the taxi driver I met outside New York City who had a lot of taxi drivers right next to him at the train station, offering the exact same product and competing on price. Many owners of auto dealerships are legally protected from someone opening a dealership that sells the same car right next to them. This makes owning auto dealerships far better than owning a taxi cab.

Law is also a big part of why beer distribution is a great business. In every state except Washington, beer distributors are protected by a three-tiered system set up after Prohibition that separates manufacturers, distributors, and retailers. Beer companies are legally prohibited from distributing beer themselves. Also, in many states only one distributor can service a region.

Laws are a surefire way to limit price competition. But there are other ways to limit it, as well. Another way is via scale. Here's how that works.

Suppose you have a product that is really, really, really hard to produce. However, once you produce it, you can reproduce it very cheaply. Now, it may be very difficult for someone else to enter your industry—and undercut you on price.

Investing and market research have this quality. Figuring out the right investments or developing a deep understanding of an industry is complex. But, once you do it, you can easily scale your product, by increasing the amount of investment or selling your research to multiple firms in an industry.

Suppose you want to enter the market research business, perhaps the most reasonable path to wealth discovered in the data.

Also suppose you have built, over your lifetime, a unique expertise in one market. This expertise has taken years to build, as it involves numerous close, long-term relationships in the industry and carefully collected proprietary data.

Now you write up reports and sell them for $5,000 per month to a lot of firms in the industry. A competitor can't easily come in and produce your same reports and sell them for a lower price. The competitor would struggle to build your same contacts and datasets. You have a moat around your business.

Another path to avoiding price competition is by building a brand that people love. This is what independent creatives can do. Artists have fans who are willing to pay extra for work produced by their favorite artist. For Bruce Springsteen's fans, a Springsteen concert is not just a generic concert. Even if another artist charged a lower price for tickets, his fans would still choose the Springsteen concert. And this is true as well for most of the roughly ten thousand rich independent creatives. They are not in commoditized businesses, in which customers will merely pick whoever charges the lowest price. They have fans who will pay more for their work.

THE ESCAPE FROM GLOBAL BEHEMOTHS

Businesses in the Big Six all offer a way for firms to have some protection against price competition, which allows them to protect their profits. However, not all industries that allow firms some way to avoid price competition are in the Big Six.

Some businesses that give firms a way to avoid price competition become dominated by one or two enormous firms, making it difficult for anyone else to compete. Consider, say, the sneakers industry. A sneaker company can be protected against price competition due to having a brand that people value. Many people will pay more for a Nike sneaker than the identical sneaker that isn't made by Nike.

However, the sneaker business fails to crack the Big Six. And data from the Tax Data Researchers shows that very few sneaker companies have rich owners. The reason that the sneaker business mints few wealthy owners is that, while it offers a way to avoid price competition via powerful brands, this advantage is captured by a few enormous firms. Nike, Reebok, and a small number of other companies can afford to pay the best athletes to advertise their products and create the largest brand advantage.

The tech industry similarly offers ways for firms to avoid price competition but it tends to be dominated by a small number of enormous firms. Operating systems and software are extremely complex to design and can be cheaply reproduced afterward, setting up a natural moat. However, there is a tendency for global behemoths, such as Microsoft, to hire the best people, design the best software, and pay the most for advertising. It is difficult for any small firm to compete with tech behemoths.

When I think about businesses in the Big Six, I realize that all of them have some natural factors that prevent domination from a few massive firms.

Think about it. Real estate markets are localized. It would be impossible for a global real estate behemoth to know everything about every local market—plus have contacts with every local politician.

The investment industry and market research industry are naturally fragmented. Investment firms have particular investment strategies that they have specialized in. Market research firms know a lot about a particular market. A global behemoth would be unable to compete with the firms' specialized knowledge.

Auto dealerships have legal protection against global behemoths. So do wholesale beer distributors. Middlemen frequently have personal relationships with local retailers that would prevent a huge firm from competing in their local market. As for art, fans of a particular artist will prefer that particular artist's work to the work of the most popular artists in the world.

HOW FAR MONOPOLIES REACH

Nowhere	To a Local Market	To a Global Market
You will be stuck in ruthless competition.	You have a decent chance of becoming rich.	You will almost certainly be defeated by a global behemoth.

SO, CAN YOU USE THIS DATA TO BECOME RICH YOURSELF? THERE
are some seemingly data-driven lessons that may be impossible
to carry out in reality. For example, you might see this data
and decide your best path to becoming rich is to own an auto
dealership. But then you might find that the people who currently own auto dealerships have zero interest in selling their
auto dealerships to you.

That said, I think you can use the spirit of the data to understand the questions you must ask about your career if you
want to get rich. I would say that there are three big, relevant
questions here.

The Get-Rich Checklist

1. Do I own a business?
2. Does the business have a path to avoid ruthless price
   competition?
3. Does the business have a path to avoid being
   dominated by a global behemoth?

If the answer to any of the three questions is no, you are
unlikely to become rich.

Of course, getting in a situation in which all three answers
are yes is not easy, which is not surprising since many people
want to be rich. Thankfully, Chapter 9 will show you that
money is far from necessary to be happy. In fact, the things
that tend to make people happy—such as gardening and walking with a friend by a lake—are surprisingly cheap and easy. A
nerdy way to say this: the Get-Happy Checklist is much easier
than the Get-Rich Checklist.

And, truth be told, what I have shown you is only the start

of the Get-Rich Checklist. There is more that goes into succeeding in business; qualities of the entrepreneur can dramatically change the odds of success. Data has uncovered that, as well.

UP NEXT

The field an entrepreneur enters is a big factor in their success. But it is not the only factor. Even in the best fields, some succeed and others don't. What determines, within a field, who succeeds? Data scientists have mined recently collected datasets on the entire universe of entrepreneurs—and found some surprising predictors of success.

CHAPTER 5

THE LONG, BORING SLOG OF SUCCESS

Every aspiring entrepreneur should hang a poster of Tony Fadell on their wall.

One day, a short while ago, Fadell was frustrated by the clunky thermostats available in his house. So, like so many entrepreneurs before him, Fadell harnessed new technology to fix a problem that frustrated him (and millions of others).

He created a company (Nest Labs) to develop a new programmable thermostat. The thermostat was sensor-based, Wi-Fi enabled, app-connected, and wildly popular.

Within a short time, the company made Fadell, like so many other tech entrepreneurs, exceedingly rich. A mere four years after starting his company, Fadell sold it to Google for $3.2 billion in cash.

There are some important points about Fadell's story that make it such a valuable one for entrepreneurs—and motivate my "Hang a Tony Fadell Poster on Your Wall" campaign. The data tells us that many aspects of Fadell's story are common among successful entrepreneurs, even if they might go against some conventional wisdom.

First was Fadell's age. When Fadell started Nest, he was not some whiz kid. He didn't start Nest from his dorm room. Instead, Fadell started the company when he was in his early forties.

Here's another important point: Fadell had already established himself as a well-regarded employee before he created his successful business. Fadell was not some serial entrepreneur who was a born risk-taker and incapable of having a boss; nor was he some career failure taking his last swing at career success. By the time Fadell started Nest, he had worked as a diagnostics engineer at General Magic, a director of engineering at Philips Electronics, and a senior vice president at Apple. By the time he started Nest, in other words, he had among the best employee résumés in Silicon Valley.

And crucially, this decade-plus of experience rising as an employee at blue-chip companies gave Fadell deeply relevant and specific skills related to the business he created on his own. When he had his the-world-needs-a-less-clunky-thermostat epiphany, he had experiences that directly prepared him to follow through on his idea.

Fadell utilized the lessons in product design that he had learned at General Magic; the lessons in managing teams and financing that he had picked up at Philips Electronics; and the lessons in improving the entire customer experience that he had absorbed at Apple. He recruited a team from the network that he had built and poured in capital that he had earned from his time as an employee at all three companies.

He had also learned from the many mistakes he had made during his twenties and thirties. In an interview on *The Tim*

Ferriss Show, Fadell said, "I wanted to look as horrible as I could in my 20s, so I could only look better as I got older." He noted that, when he first was put in charge of other people at Philips, he was "probably the worst manager under the sun." He would, for example, arrogantly lecture his underlings. But, by listening to feedback, he realized the importance of empathy for leadership; by seeing situations from other people's point of view, he came to realize, he could more easily persuade them to pursue the best course of action. These lessons were learned—and Fadell's management skills well honed— by the time he had to lead his team at Nest.

In the previous chapter, we discussed the revelations from tax records that show that entrepreneurship is the best path to wealth and clarify which fields might be most fertile for an entrepreneur. But new, big datasets also tell us that, whatever field you are in, certain career decisions make you more likely to be among the entrepreneurial successes in that field. There is a clear formula for increasing the likelihood of entrepreneurial success. It is to follow a path like that of Fadell: spend many years building expertise and a network while proving your success in a field before striking out on your own in middle age. In fact, the new data busts some myths about entrepreneurs.

MYTH: THE ADVANTAGE OF YOUTH

Think of a successful founder of a business. Who is the first person who comes to mind?

Unless you have spent the past few minutes Googling around for how to create a Tony Fadell poster, it is probably someone like Steve Jobs. Or Bill Gates. Or Mark Zuckerberg. One feature these world-famous founders all have in common is the stage of life at which they founded their business empires. They were all young. Jobs started Apple when he was twenty-one years old. Gates began Microsoft at nineteen. Zuckerberg created Facebook at the age of nineteen.

It is not a coincidence that so many young people come to mind when we think of successful entrepreneurs. The media, when they write of entrepreneurs, tend to focus on younguns. A recent study examined all entrepreneurs who were featured in the "Entrepreneurs to Watch" section of two prominent business magazines. The median age of these featured founders was twenty-seven years old—a tad older than Jobs, Gates, and Zuckerberg but still well short of middle age.

Venture capitalists and investors have bought into the media-driven narrative that younger people are more likely to build great companies. Vinod Khosla, a cofounder of Sun Microsystems and venture capitalist, said, "People under 35 are the people who make change happen . . . people over 45 basically die in terms of new ideas." Paul Graham, the founder of Y Combinator, the famous start-up accelerator, said that, when a founder is over the age of thirty-two, investors "start to be a little skeptical." Zuckerberg himself famously said, with his characteristic absence of tact, "Young people are just smarter."

But, it turns out, when it comes to age, the entrepreneurs we learn about in the media are not representative. In a pathbreaking study, a team of economists—Pierre Azoulay, Benjamin F. Jones, J. Daniel Kim, and Javier Miranda (henceforth referred to as AJKM)—analyzed the age of the founder of every business created in the United States between the years

2007 and 2014. Their study included some 2.7 million entrepreneurs, a far broader and more representative sample than the dozens featured in business magazines.

The researchers found that the average age of a business founder in the United States is 41.9 years old—in other words, more than a decade older than the average age of founders featured in the media. And older people don't just start businesses more than many of us realize; they also succeed at creating highly profitable businesses more often than their younger peers do. AJKM used various metrics of success for a business, including staying in business for longer and ranking among the top firms in revenue and employees. They discovered that older founders consistently had higher probabilities of success, at least until the age of sixty.

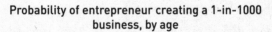

Probability of entrepreneur creating a 1-in-1000 business, by age

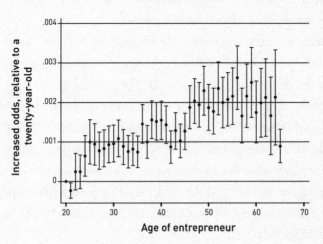

Source: Azoulay et al. (2020)

A sixty-year-old start-up founder has a roughly three times higher chance of creating a valuable business than a thirty-year-old start-up founder.

Further, the surprising—and largely hidden—success of older founders is true in every field that AJKM studied, even in tech, perhaps the field most associated in people's minds with young founders. The average age of a founder of a highly profitable technology company, the researchers found, is 42.3 years old. Data reveals, in other words, that the media-driven idea that successful founders tend to be young is plainly false.

The idea that young people are more likely to succeed in business is not merely a media-generated myth. It is a dangerous media-generated myth. In 2010, *The Social Network*, written by Aaron Sorkin, famously popularized the story of one of the nonrepresentative young, successful entrepreneurs: Zuckerberg. The movie, which told the story of how Zuckerberg created Facebook from his dorm room, earned legions of fans, grossing more than $200 million.

Some of those fans were encouraged to copy the hero of the movie. A study found that in the years following the release of the movie, teenage entrepreneurship rose eightfold. This is despite the fact that teenage entrepreneurship, no matter what *The Social Network* implies, has always been an awful bet. Or, as I like to say, the plural of *The Social Network* isn't data.

So, why are successful entrepreneurs more often than not in the second half of their lives? Well, they tend to have spent many years learning about a business. This relates to the next myth about successful entrepreneurs.

MYTH: THE OUTSIDER'S EDGE

Suzy Batiz is among America's richest self-made women. Her path to success involved essential oils, trial and error, and feces. In 2007, she invented Poo-Pourri, a product that, when you spray it in a toilet, makes one's defecations less smelly.

Batiz came up with her "aha" moment during a dinner party conversation about bathroom odors. (One might say: *Just as an apple hit Newton on the head, motivating him to come up with his theory of gravity, the stench of feces hit Batiz on the nose, motivating her to come up with Poo-Pourri.*) Shortly after having her insight, Batiz got to work. She experimented with various essential oils until at last she stumbled on a mixture that suspended perfectly on the surface of water in a toilet bowl. She followed family members as they sampled her product, smelled the results, and concluded she had a winner.

She did. Buoyed by a viral commercial featuring a pretty redhead sharing potty humor and a rave review on the *Today* show, the product took off. Batiz's net worth is now estimated to be more than $200 million. (One might say: *Batiz went from a "conversation about shit" to "fuck you money" in less than a decade.*)

What stands out about Batiz's path to riches is not just the field: feces odor reduction. What is also striking is how little relevant experience she had. At the time she began testing her essential oils, Batiz had no training in chemistry and no experience in consumer goods. She also had no career success. She had declared bankruptcy after a bridal salon she had purchased with her husband failed. She also attempted a clothing line, which failed; a hot tub repair business, which failed; and a tanning salon, which also failed.

Could Batiz's be a common pattern of success?

David Epstein, in his best-selling book *Range*, has a chapter called "The Outsider Advantage" that makes a seemingly counterintuitive claim: outsiders frequently have an edge in solving difficult problems. Epstein notes the many difficult problems that had stymied an entire field only to be solved by somebody outside that field. For example, in the early eighteenth century, one of the most important open problems in chemistry was to discover a substance that could reliably preserve food. Many of the greatest minds in the world, including Robert Boyle, the "father of modern chemistry," had failed. A solution finally came from the confectioner Nicolas Appert: heating champagne bottles in boiling water. Appert started a successful business based on his invention.

Epstein notes that people within a field frequently only know how to try methods that have worked before. But innovation frequently requires new methods, methods that outsiders might be more likely to think up and try. As Epstein put it, "Sometimes, the home field can be so constrained that a curious outsider is truly the only one who can see the solution."

Could this extraordinary and provocative theory be true? In entrepreneurship, can there be an outsiders' edge? Are stories such as that of Suzy Batiz common? If you want to create a business, should you be drawn to fields outside your area of expertise, where your lack of experience may give you a leg up against insiders who are so "constrained" by their "home field"?

No. Once again, Big Data definitively rejects this idea.

AJKM, in addition to studying the age of entrepreneurs, studied the employment history of entrepreneurs. In particular, over their entire sample of founders, they looked at whether founders had previous job experience in the same industry that they founded their company in. For example, did someone who started a soap manufacturing business previously work at a company that focused on manufacturing soaps? The researchers also looked at how successful the business was: for example, did it reach among the top 1-in-1,000 businesses in that field in revenues?

The researchers found that there is an enormous "Insider's Advantage" in entrepreneurship. Entrepreneurs are roughly twice as likely to build an extremely successful company if they previously worked in the field in which they started the company. The advantage gets larger the more directly the previous experience is related to the business. Someone who has already worked in a narrow field—such as manufacturing soaps—is more likely to succeed in a soap-manufacturing business than someone who worked in an adjacent field, such as manufacturing food.

In business, deep domain knowledge is not a curse, preventing entrepreneurs from recognizing innovative opportuni-

THE INSIDER'S EDGE IN ENTREPRENEURSHIP

Work Experience of the Founder	Probability Business Will Be a 1-in-1,000 Success
No experience in field	0.11%
Experience in same broad field—but not narrow field	0.22%
Experience in same narrow field	0.26%

Source: Azoulay et al. (2020)

ties. In business, being the consummate insider is, on average, a large edge.

MYTH: THE POWER OF THE MARGINAL

Suzy Batiz wasn't just an outsider to the field of fecal-fighting chemistry when she began her business. She was also, by just about any conventional metric, a failure. Recall that she had declared bankruptcy and had numerous business ventures go bust. She was outside the margins of the successful.

Could Batiz's lack of success actually, as strange as it may seem, have been an advantage?

Paul Graham, the brilliant essayist and founder of Y Combinator, a start-up accelerator, wrote a fascinating and provocative essay arguing that people who have failed a lot can actually have an edge in entrepreneurship. In the essay, called "The Power of the Marginal," Graham notes that "great new things often come from the margins."

Graham points to the examples of the founders of Apple, Steve Jobs and Steve Wozniak. Jobs and Wozniak, Graham writes, "can't have looked good on paper" when they started their now-iconic company. At the time, they were "a pair of college dropouts" and "hippies" whose only business experience consisted of creating blue boxes to hack into phone systems.

Graham suspects that stories such as that of Jobs and Wozniak—successful founders who didn't look good on paper—may not be mere anomalies. He suspects that marginal people may have a surprising edge in business. He has very clever theories for why being a less successful person may actually be an advantage. For example, Graham notes that insiders can be "weighed down by their eminence," causing them to avoid any risks. Those on the margin, in contrast, have nothing to lose and can risk everything.

So, is this provocative and seemingly counterintuitive theory of entrepreneurial success true?

No. "The Power of the Marginal," like the power of youth and outside status, is a myth.

The Tax Data Researchers examined the previous wage histories of every start-up founder in the United States—and crossed this with data on the profitability of their businesses.

They could examine whether, as Graham's theory suggests, people who had the greatest success prior to starting their business might struggle as entrepreneurs. The authors found this was not the case. Inconsistent with Graham's idea, the conventionally successful massively outperform other employees as entrepreneurs.

As shown in the chart on the next page, the odds of success of a business venture was highest when the founder had earned in the top 0.1 percent of salaries in their field before they started their business: hardly marginal people; hardly people who didn't have any eminence to protect.

THE COUNTER-COUNTERINTUITIVE IDEA

Let's be honest. When you step back and think about the data findings in this chapter, they aren't all that surprising.

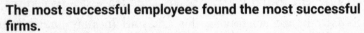

The most successful employees found the most successful firms.

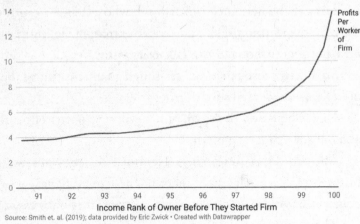

Income Rank of Owner Before They Started Firm

Source: Smith et. al. (2019); data provided by Eric Zwick • Created with Datawrapper

Data scientists have mined large, new datasets to discover that entrepreneurs are more likely to succeed after having spent many years rising to the top of their field. Isn't that intuitive? Shouldn't rising to the top of a field be positively associated with entrepreneurial success in that field?

Yet, some of these findings, as intuitive as they may seem, go against narratives that have captured the imagination of the public. Many people have been captured by the three myths debunked by data: the Advantage of Youth, the Outsider's Edge, and the Power of the Marginal. It turns out that the findings in this chapter fall into my favorite bucket of ideas: counter-counterintuitive ideas. Here's how counter-counterin-tuitive ideas work:

There is initially a commonsense idea—say, that being a bit older and wiser would be helpful in starting a business. But then there are some people whose life story goes against this commonsense idea. Some extremely young people achieve enormous business success—say, Mark Zuckerberg. Because these people's life stories go against common sense, they make for great stories—"Who would think that a nineteen-year-old could create a multibillion-dollar business?"

People love telling the surprising stories—or making movies about them. Aaron Sorkin wrote *The Social Network* about nineteen-year-old Mark Zuckerberg and not *The Thermostat* about forty-one-year-old Tony Fadell.

So many people hear so many of these initially surprising stories. And the stories, which initially got traction in part because they were surprising, now begin to feel natural. People think that youth can be an advantage in entrepreneurship. "Just think of *The Social Network*," they think. The initially surprising idea, because it is so striking and the stories are told so frequently, now is considered conventional wisdom.

Representative, large datasets have no bias in favor of examples that make for interesting stories and sometimes tell us that counterintuitive ideas that have become conventional wisdom are incorrect. When you look at the entire universe of entrepreneurs—not just the universe of entrepreneurs whose stories are most frequently told—you see that age and wisdom are advantages in creating a successful start-up.

You may now be wondering what my other favorite counter-counterintuitive ideas are. Or you may not. But regardless, here are my favorite examples of Big Datasets countering a popular counterintuitive idea and returning us to the intuitive idea:

» **NBA players are more likely to come from middle-class, two-parent backgrounds.** Some NBA players come from low-income backgrounds. Precisely because it is so striking to see someone born in poverty have such enormous success, these stories are told more. This encouraged some to believe that coming from a rough background gave basketball players more motivation to work hard and reach the NBA (the counterintuitive idea). For example, in the book *The Last Shot*, a college coach questions whether a suburban athlete is "hungry enough" to succeed. However, research from Joshua Kjerulf Dubrow and Jimi Adams, as well as my independent research, has found that NBA players disproportionately come from middle-class families.

» **Jokes are more likely to be looked for when people are happy than sad.** It is very striking when someone tells jokes during a tragedy. Because the contrast between a dark time and jokes is so striking, we notice jokes told during dark times more. This led some people to the idea that pain could be a bigger motivator than pleasure to laugh (the counterintuitive idea). As Charlie Chaplin put it, "Laughter is the tonic, the relief, the surcease from pain." But I studied Google searches for jokes. Searches for jokes are lowest on Monday (the most miserable day of the week), lower on freezing days, and they plummet after major tragedies, such as the Boston Marathon bombing. The data tells us that people are more likely to laugh when things are going well (the counter-counterintuitive idea).

» **Greater and greater intelligence is an advantage in life.** Lots of people mess up their lives. But it is really striking when an extremely smart person screws up. The salience of these examples led to theories that people can be "too smart for their own good." Popular books such as *The Intelligence Trap* and *The Intelligence Paradox* argued that too much intelligence becomes a disadvantage (a counterintuitive idea). But a recent study of tens of thousands of people found that IQ is an advantage in just about every area of life, and there is no point on the IQ distribution at which more IQ stops being an advantage. The data tells us that more intelligence is always an advantage (the counter-counterintuitive idea).

TRUST IN THE DATA: THE PATIENT PATH TO SUCCESS

Data clears away the noise that comes from the media's un-representative examples in understanding entrepreneurial success.

When you get rid of all the noise—the stories we hear

from the media or from friends and acquaintances—and look at the actual data on entrepreneurial success, there is a formula for maximizing your chances of leading a successful business. The formula is this: spend many years learning the ins and outs of a field, prove your worth in that field by being among the best-paid employees, and then go out on your own to make your real fortune.

This success formula isn't necessarily the most exciting. It might be more exciting to think that you are ready, in your early twenties, having mastered few skills, to create a business empire. It might be more exciting to think that the best path to business success is to mix a few essential oils you know little about and become rich shortly afterward. It might be exciting to think you are ready to strike out on your own even though you have not proven any success in your field. It might be more exciting to think that you don't really need to know much about your field, that you will learn all that is necessary as you create your business. But these titillating ideas are false. They are the myths of success, not the data science of success.

The success formula also isn't necessarily easy. It requires tremendous discipline. If you follow the data-driven formula of entrepreneurial success—devoting your twenties and thirties to mastering skills and proving your worth in a narrow field—you will undoubtedly, during your period of achieving mastery, learn of others in your age cohort who have already had enormous entrepreneurial successes. You will undoubtedly learn of some overnight successes.

While most business successes come in middle age, some famous outliers do achieve outsized success earlier. While most business successes require mastery of a field, some tiny few do strike gold in a field they know little about. While most business successes require decades of hard work and mastery of a field, a small number of entrepreneurs do just get lucky. These one-off stories will not be representative of success. They will, in fact, be deeply misleading about the best path to success. But they will, no doubt, make it difficult to keep grinding away.

When you hear of these stories, you might look again at the charts presented in this chapter. Heck, if you are really nerdy, you might print them out and hang them on your wall, next to your Tony Fadell poster. Glance at the charts; glance at the Fadell poster. Then get back to work.

Trust in the data!

UP NEXT

One can have a higher chance of being successful if one patiently builds skills in a narrow field—and then strikes on their own. But, let's be honest, luck plays a large role in success. Big Data, including deep dives into the sales of hundreds of thousands of artists, can tell us how luck works. And you can use data-driven insights to become luckier.

CHAPTER 6

HACKING LUCK TO YOUR ADVANTAGE

In October 2007, Brian Chesky and Joe Gebbia, two room-mates living in San Francisco who had met in art school and were now unemployed, came up with an idea to help them pay their bills. A major design conference was coming to San Francisco, and, as hotels were likely to be fully booked, Chesky and Gebbia could rent the spare air mattresses in their apartment and serve breakfast to guests who couldn't find a place to stay.

Some conference attendees indeed rented these air mattresses. And Chesky and Gebbia, two men never short on confidence and long harboring entrepreneurial ambition, decided they might have stumbled on their big idea. Perhaps people around the world could similarly earn money by renting out spare air mattresses and serving breakfast. Chesky and Gebbia brought in a mutual friend and computer whiz, Nathan Blecharczyk, and they had a website for their idea: airbedandbreakfast.com.

In the ensuing weeks and months, Chesky's and Gebbia's

idea went absolutely nowhere. A few people wanted to rent out their spare air mattresses. A few others yearned for an air mattress to sleep on. But these numbers were not enough to make a real business. Soon Chesky and Gebbia had each accumulated more than $20,000 in credit card debt. And Blecharczyk, the only member of the operation capable of coding, gave up on the project and moved to Boston.

Chesky and Gebbia, never lacking in scrappiness, took a couple of trips in hopes of saving their business. They went to the South by Southwest conference in Austin, thinking that the enormous number of guests in town might give their business a big boost. It didn't. The two young men, however, did meet and befriend a well-connected man in Silicon Valley named Michael Seibel.

Chesky and Gebbia went to Denver for the 2008 Democratic convention, thinking that the enormous number of guests in town might give their business a big boost. It didn't. The two young men, however, did design and sell cereal based on the presidential candidates—Obama O's ("The breakfast of change") and Cap'n McCain's ("A maverick in every bite"). Remarkably, they sold enough to pay back their debt.

In any case, their business was basically dead when, one evening, Chesky and Gebbia met up with Seibel, the man who had liked them in Austin. Seibel, still impressed with the two young men, suggested that they apply for Y Combinator, a start-up accelerator that was run by his friend Paul Graham in Silicon Valley. The deadline had passed, but Seibel had enough pull with Graham to get their application a look. This was their first big break.

Graham didn't like Chesky's and Gebbia's business idea,

but, when told of their cereal story, he was impressed with their moxie. Graham gave them $20,000 in seed funding, which was enough to convince the technical cofounder Blecharczyk to rejoin and allowed the team to survive a few months longer.

Soon the founders would get their next big break. David Rozenblatt, a drummer for Barry Manilow, who was about to go on tour and had heard of Chesky's and Gebbia's site, requested that he be allowed to rent out his whole apartment—beds and all. The airbedandbreakfast.com founders initially said no, noting he would be unable to provide breakfast if he wasn't there.

But Manilow's drummer's request did cause Chesky and Gebbia to step back and think and soon have their proverbial "aha" moment. The two entrepreneurs realized that there was a much bigger business that was a cousin of their initial idea: people renting apartments when they were gone.

Drop the air mattresses. Drop the breakfast. And allow millions of people, around the world, when they are out of town—to drum for Barry Manilow or for any other reason—to make some extra cash by renting out their place.

Airbedandbreakfast.com was rebranded as Airbnb—and immediately began to get traction. Sure, few people wanted to set up an air mattress in their apartment and serve breakfast to out-of-town visitors. But it turned out that millions of people around the world wanted to rent out their empty places. (The full Airbnb story is told in Leigh Gallagher's excellent book *The Airbnb Story*.)

There was one remaining problem: the Airbnb team needed money to continue their business. They were in the midst of the Great Recession; investors around the world had tightened their belts. Further, many investors told them the hospitality market was not large enough to make it worth their time.

Soon Chesky and Gebbia would get their final big break. One day, Greg McAdoo, a partner at Sequoia Capital and an old friend of Paul Graham, came by the Y Combinator office. McAdoo, unlike most investors, was convinced that now was a good time to invest in companies—as others would be scared off. He also had a theory that scrappy people were the ones most likely to build a company during an economic downturn. Remarkably, he also just happened to have spent the past year and a half analyzing the vacation rental market. He had determined that the market was actually worth $40 billion, far more than what others thought. He met the Airbnb team and was immediately ready to send a check for $585,000. Airbnb now had a product people wanted and the money to get their idea off the ground. They were now on the path to a multibillion-dollar valuation.

As Tad Friend wrote in the *New Yorker,* Airbnb's rise "seems replete with luck." There was the lucky meeting of Seibel in Austin. There was the lucky encounter with the perfect investor, McAdoo, at Y Combinator. And of course, there was Manilow's drummer. Had Manilow not been going on tour at just the right moment, Chesky and Gebbia might have never discovered their right business model before they went broke. Some other group of entrepreneurs may have been the ones to build this business years later.

The difference between being a billionaire and a lifelong struggling entrepreneur can come down to an idea from Barry

Manilow's drummer. Sometimes, just when you are about to go broke, Barry Manilow goes on tour and puts you on the path to becoming a billionaire.

Sam Altman, who replaced Paul Graham as CEO of Y Combinator, has watched thousands of start-ups succeed or fail—and built a model in his head of what it takes to win in Silicon Valley. In a 2014 lecture at Stanford, Altman summed up the formula for entrepreneurial success as follows: "something like Idea times Product times Execution times Team times Luck, where Luck is a random number between zero and ten thousand."

Thousands of unknown entrepreneurs or actors may have drawn a 1,000 or a 500 or a 0 as their luck number. Chesky and Gebbia seemingly drew a 10,000.

IT IS COMMON IN SOME CIRCLES TO SPEAK ABOUT THE OUTSIZED role that luck plays in success. Many successful people attribute a good portion of their success to luck. The Nobel Prize–winning economist and *New York Times* columnist Paul Krugman says of his success: "I was very lucky to be in the right place at the right time." John Travolta, the actor, explained the reason for his success: "I got lucky." Ditto the actor Anthony Hopkins, who said, "I think I've been very lucky."

But is it possible that we are exaggerating the role of luck in life? Some interesting data suggests that luck may play a smaller role in life than some of us think. A wide range of studies have found fascinating patterns of behavior that consistently lead to seeming good luck.

One of the most important early, data-driven studies on luck was conducted by the business researchers Jim Collins and Morten T. Hansen. While their study focused on how luck influences large companies, their results have implications for understanding luck in every field.

Collins and Hansen first assembled a list of what they called 10X companies—some of the most exceptional companies in history. To qualify for the 10X company club, a business had to outperform its peers in the stock market by a factor of at least 10 for an extended period of time. The companies that qualified included Amgen from 1980 to 2002, Intel from 1968 to 2002, and Progressive Insurance from 1965 to 2002.

Next, for each 10X company, the researchers found a comparison company that was in the same industry and started at a similar size but never outperformed its peers. Amgen's comparison was Genentech; Intel's comparison was AMD; Progressive's comparison was Safeco.

The researchers then pored through any document they could find on the history of the 10X companies and the comparison companies to find what they called "luck events." They wanted to see how many more lucky breaks 10X companies had than comparison companies.

They defined a luck event as having three properties:

1. "some significant aspect of the event occurs largely or entirely independent of the actions of the key actors in the enterprise,"
2. "the event has a potentially significant consequence (good or bad)," and
3. "the event has some element of unpredictability."

The authors indeed found many lucky events in the 10X companies. Each 10X company, the researchers discovered, had an average of seven events that were entirely outside the company's control that significantly improved the business.

For example, in studying Amgen, the authors found that an enormous part of their success was due to a Taiwanese scientist named Fu-Kuen Lin, who just happened to locate and respond to a small help-wanted classified advertisement by Amgen. Lin turned out to be a perseverant genius who worked tirelessly to discover the genetic blueprint for erythropoietin, a protein that helps produce red blood cells in kidneys. Lin's work led to the discovery of Epogen, one of the most profitable drugs in biotech history. Had Lin not stumbled on that ad, it is likely Amgen would have never discovered Epogen. It is easy to imagine a different history for Amgen where Lin doesn't see that ad, Epogen is never discovered, and Amgen is not a 10X company.

Amgen seemingly got lucky. It would be easy for Amgen to think that it was unusually lucky. It would be easy for Amgen's competitors to note the Lin break and say something like, "Well, Amgen stumbled on Lin. We weren't so lucky."

And if Collins and Hansen had just studied the 10X companies, they might have concluded that every successful company had many extremely lucky breaks. But the researchers didn't just look at the history of the 10X companies. They also looked at the history of the comparison companies.

Even though these businesses never outperformed their peers, Collins and Hansen found, they also had plenty of lucky breaks during their histories. Genentech, for example, barely beat other companies in a race to be the first to use gene splicing to create a human insulin approved by the Food and Drug Administration. Had their work been even slightly delayed, another company would have likely beaten them to that profitable market. In fact, Collins and Hansen found that Amgen and Genentech had roughly the same number of big breaks.

Here's the remarkable result from Collins and Hansen. Across the companies in a range of fields, there was no statistically significant difference in the number of lucky breaks received by 10X companies and 1X companies. The 10X companies averaged about 7 lucky breaks. The 1X companies averaged about 8 lucky breaks.

Collins and Hansen concluded that successful companies didn't have more luck; they were better able to capitalize on the luck that they got, the luck that any company can expect.

Collins and Hansen make an extremely important point. Just about everybody, over the course of a lifetime, can expect some fortuitous opportunities. Imagine a person who never once met a person who could help them out in life, was never paired up with someone unusually talented, and never met someone who needed their skill set. This person would legitimately be the unluckiest person of all time. An average amount of luck in life includes many seemingly fortuitous opportunities. People or organizations who are more successful recognize these fortuitous opportunities and capitalize on them.

Return to Airbnb, the story that supposedly shows how much luck matters in businesses. Sure, Airbnb had some fortuitous opportunities. But they also utilized the luck that they were handed. How many unsuccessful companies didn't sell cereal to make money when they were broke? How many unsuccessful companies didn't network so they would have connections to save their company? How many unsuccessful companies didn't apply to a start-up accelerator when they knew they needed a boost? How many unsuccessful companies didn't pivot to a new idea when they realized their current one wasn't working?

Airbnb didn't necessarily get unusually lucky. They capitalized on the luck that anybody who works hard can expect. And, as lucky as Airbnb had historically been or seemed to be, they got an unusually bad luck draw when a global pandemic stopped people from traveling.

Shortly after the pandemic started, Airbnb's bookings dropped 72 percent; its estimated valuation fell from $31 billion to $18 billion; and the company was forced to put their IPO on hold. But, like all extremely successful businesses, Airbnb was good at working around their unlucky breaks. They quickly cut costs and pivoted to more of a focus on long-term stays. They also gave unusually generous severance packages to laid-off workers and refunds to guests, which generated positive press. Instead of whining about how unfair it was that a pandemic hit just before they were to launch an IPO, Airbnb worked hard around the problem. The company posted a surprise profit by the end of 2020 and IPO'd at a more than $100 billion valuation.

Collins and Hansen's study suggests that, whereas we see the lucky breaks of successful people or organizations, there was actually good decision-making underlying that luck. This suggests that successful people or organizations do things that make them seem luckier. Indeed, new studies, many of them focusing on artistic success, have uncovered some of these luck-creating strategies. Or, as I sum up some of the fascinating new work from data science: there are patterns underlying luck.

THE ART WORLD AS A PLACE TO LEARN HOW TO HACK LUCK

In his brilliant book *The Formula*, Albert-László Barabási, a physicist at Northeastern University who has explored the mathematical patterns underlying success, notes that fields differ in how much performance can be measured. In particular, Barabási points out a clear difference between sports and art.

In sports, the quality of a player is reasonably easy to establish. Michael Jordan in his prime was clearly the best basketball player in the world, scoring more points and leading his team to more championships than everybody else. Michael Phelps in his prime clearly swam faster than everybody else. Usain Bolt clearly ran faster than everybody else.

Sports nerds such as myself might decide to skip socializing in high school to try to develop new statistics to determine whether the baseball player that conventional wisdom says is 100th best might actually be 86th best. But we generally all agree on the difference between a Major League Baseball player and a minor-league player. The best athletes in the world are, by and large, discovered and given a shot.

This isn't true for artists. In art, quality is harder to assess. Laypeople and even art critics sometimes have a difficult time judging art. The *Washington Post* columnist Gene Weingarten convinced the world-renowned violinist Joshua Bell to play as a busker at a Metro station in Washington, D.C. Only 7 of the 1,097 people who passed by stopped to listen. Another enterprising journalist, Åke "Dacke" Axelsson, got a four-year-old chimpanzee to paint "modern" art and found many art critics who praised the work.

You know that annoying, immature guy who looks at a painting at a museum, particularly a modern work, and goes "I don't see why that's so special"? That annoying, immature guy is me, and that annoying, immature guy is, studies suggest, often justified.

There are two prominent effects in a world in which performance is hard to judge.

THE MONA LISA EFFECT: THE MOST POPULAR WORK GOT LUCKY

The first is what I call the Mona Lisa Effect, which, as you can probably guess, I named after the painting the *Mona Lisa*. The Mona Lisa Effect says that unpredictable events massively influence success.*

It turns out that an unpredictable event turned the *Mona Lisa* into the world's most famous painting. You might think that the *Mona Lisa* is the most famous painting in the world because of the qualities of the work: the eyes of the subject (they seem to stare back at you from wherever you are looking); that mysterious half smile; the features of the subject

---

* *There is another Mona Lisa Effect based on the idea that the subject's gaze follows you no matter where you stand.*

(the high forehead and pointed chin, the average woman that seems so easy to fall in love with).

But, the truth is, for most of its first 114 years hanging in the Louvre, the *Mona Lisa*—with the same eyes, smile, and face—was just one of many great paintings. Day after day, it hung on the walls of the Louvre and did not stand out among all the other world-class artwork in the museum.

Time for the first true-crime story of this book!

On one Tuesday morning in the late summer of 1911, a guard walked into the Louvre and noticed that the *Mona Lisa* was gone. All that were left were four hooks that had been used to hold the piece.

By the evening, a special edition of *Le Temps,* the major morning newspaper in France, was printed to tell the story. The next day, the missing *Mona Lisa* was the top story in newspapers around the world.

If people hadn't known the *Mona Lisa* beforehand, they pretended they did. If people didn't feel shock, they feigned it. "What happened to the *Mona Lisa*?" became a worldwide phenomenon, with press attention rivaling that of wars.

Early on, police suspected that a young German boy may have stolen the work. The boy had visited the museum multiple times, and police thought that the boy may have become so obsessed with the woman in Da Vinci's painting, so crazy in love, that he stole the work. Remarkably, there was great sympathy for the boy, with some thought leaders at the time suggesting that a boy so in love might deserve the painting.

The investigation for a time focused on J. P. Morgan, the American banking magnate. Many people in France suspected that only an American would be brazen enough to decide that he alone deserved to enjoy the *Mona Lisa*. After it was discovered that Morgan had been vacationing in Italy at the time of the heist, the press hounded him.

For a time, the police focused their investigation on a group of artists that included Pablo Picasso. Picasso, in those days, was leading a group of young modern artists. And, after a tipster had discovered that this group had lived by a maxim that artists have "to kill one's fathers," police thought that the group might have orchestrated the heist as the ultimate murder of Renaissance art.

As with most true-crime stories, the solution was less captivating than the theories that had been concocted: it turned out that a low-level Louvre employee had stolen the piece, thinking that it might increase the value of a friend's copies of the painting. Police found the bumbling criminal two years after his crime when he attempted to sell the work to an Italian gallery.

But, anticlimax aside, the *Mona Lisa* had received unprecedented publicity over a two-year period. When the painting was returned to its rightful place in the Louvre, crowds flocked. Everybody wanted to see the work that they had been hearing so much about—the painting that was so remarkable that J. P. Morgan was thought to want it for himself or Picasso was thought to want it removed from the world. Now scores of people stopped to admire those eyes, that smile, that face.

A seemingly random event—a heist that nobody could have predicted—vaulted the *Mona Lisa* from one of thousands of well-respected paintings into the most famous painting in the world. If that heist had never happened, the *Mona Lisa* might be just another painting in one arm of the Louvre that most tourists peek at before moving on with their Paris vacation.

If that heist had never happened, the *Mona Lisa* would have been just another painting that I ignored on my 1996 family trip to Paris, as I threw a temper tantrum complaining that none of the other kids from New Jersey had to go to museums with their parents and Garrett and Mike were probably at a Mets game right at that very moment and why did I have to be in some stupid building in some stupid city in some stupid country with some stupid stuff hung on the wall. So much for the Mona Lisa Effect.

THE DA VINCI EFFECT: IT'S NOT WHAT YOU PRODUCED; IT'S WHO YOU ARE

Second, when quality is hard to judge, there is a Da Vinci Effect, which was first coined in a blog post by Jeff Alworth in 2017. The Da Vinci Effect says that the success of an artist begets more success for that artist. People are willing to pay more for the work of an artist who is already famous.

Indeed, there are many examples of pieces of art that dramatically changed in value when experts changed their mind regarding who created it. Consider, for example, the *Salvator Mundi,* a depiction of Jesus Christ. In 2005, it was sold for less than $10,000. In 2017, a mere twelve years later, it was sold for $450.3 million, the highest price ever for a piece of art. What caused the price to rise so much in such a short time? In the in-between years, art experts became convinced that the painting had been created by Leonardo da Vinci. In other words, the same painting is worth 45,000 times more just because Da Vinci drew it.

How should artists who dream of greatness work in a

world driven by the Mona Lisa Effect and the Da Vinci Effect? The predominant way that many people deal with these effects is by whining. *"Life is so unfair! That work isn't actually better than mine."*

Normally, I am all in favor of whining—and consider it my predominant coping mechanism for adult life. However, I must admit that the data definitively overrules the justification for whining. Scientists have uncovered patterns of people who tend to succeed as artists, those who have found ways to make the randomness in artistic success work for them. Moreover, unlike particular drawing or singing techniques, these techniques that artists use to get more than their fair share of luck can be used by non-artists as well.

SPRINGSTEEN'S RULE: TRAVEL WIDELY TO FIND YOUR BREAK

Barabási, the physicist who studies the mathematics of success, and a team of other scientists, led by Samuel P. Fraiberger, studied what predicts success in the art world. They worked with an app, Magnus, that collects information on the exhibitions and auctions of painters, to create one of the most remarkable datasets of artistic achievement ever assembled.

They had data on the career trajectories of 496,354 painters. For each painter, the scientists knew the vast majority of places the painter presented and the price of the vast majority of pieces that they sold.

The researchers first documented something like a Da Vinci Effect. They found that, once a painter had presented in an extremely prestigious gallery, the painter's odds for career success went up enormously. Artists who presented at one of the top art galleries—such as the Guggenheim in New York City or the Art Institute of Chicago—had a 39 percent chance to still be presenting art a decade later. More than half of these artists continued to present in high-prestige galleries for the remainder of their career. The average top price for an artwork they sold was $193,064. The prospects of artists who hadn't presented in prestigious galleries were far less rosy. Eighty-six percent of these artists were no longer working in a decade. They had an 89.8 percent chance of ending their careers remaining outside the top galleries. Their top work, on average, fetched only $40,476.

The scientists noted that, once your work has been exhibited at a major gallery, you are an insider and vouched for. Museum curators will be happy to show your work; people will be happy to buy your work. The data makes clear the easy lives of these fortunate few; more and more success and money are thrown at them.

When you hear of how advantageous it is for an already-vouched-for artist to succeed, it can be easy to get mad, to whine. "My work is better than that guy's," an outsider artist may say. "People are just buying it because he has the right credential." But this whine misses a crucial fact: most of the people inside the club started their lives outside the club. They had to do something to become a vouched-for artist whose career was on cruise control.

Here is where Fraiberger and his team's study of painters' careers gets interesting. They found there was a notable strategy that characterized artists who successfully went from outsider artists to insider artists: "a relentless and restless early search."

The researchers found that you could divide outsiders into two categories of artists. Category 1 artists presented their work in the same gallery over and over again. Category 2 artists presented their work in different galleries around the world. No, the Guggenheim wouldn't accept them. (They weren't insiders yet.) But these break-hunters found other galleries that would take them.

To see the difference between the two categories, here are examples of the exhibition schedules of one artist in each. First, this was the exhibition schedule of a Category 1 artist, an outsider who never fully broke through.

SCHEDULE OF A CATEGORY 1 ARTIST AS A YOUNG MAN

Date of Exhibition	City	Country	Institution
2004-02-13	Waitakere City	New Zealand	Corban Estate Arts Centre (CEAC)
2005-02-15	Herne Bay	New Zealand	Melanie Roger Gallery
2006-03-14	Herne Bay	New Zealand	Melanie Roger Gallery
2007-04-17	Herne Bay	New Zealand	Melanie Roger Gallery
2007-10-02	Herne Bay	New Zealand	Melanie Roger Gallery
2008-04-15	Herne Bay	New Zealand	Melanie Roger Gallery
2008-07-05	Lower Hutt	New Zealand	The Dowse Art Museum
2008-09-09	Herne Bay	New Zealand	Melanie Roger Gallery
2009-02-11	Herne Bay	New Zealand	Melanie Roger Gallery
2009-08-29	Christchurch	New Zealand	Christchurch Art Gallery Te Puna o Waiwhetu
2009-10-21	Herne Bay	New Zealand	Melanie Roger Gallery
2010-11-24	Herne Bay	New Zealand	Melanie Roger Gallery
2010-11-30	Wellington	New Zealand	Bartley and Company Art
2011-01-26	Herne Bay	New Zealand	Melanie Roger Gallery
2011-10-04	Wellington	New Zealand	Bartley and Company Art

Note that the Category 1 artist presented his art repeatedly in the same places in his home country.

Next, here is the exhibition schedule of a Category 2 artist, the German artist David Ostrowski, who did break through:

SCHEDULE OF A CATEGORY 2 ARTIST AS A YOUNG MAN

Date of Exhibition	City	Country	Institution
2005-10-19	Cologne	Germany	Raum für Kunst und Musik e.V.
2005-11-13	Eupen	Belgium	IKOB—Museum für Zeitgenössische Kunst Eupen
2006-10-20	Culver City	United States	Fette's Gallery
2006-10-25	Cologne	Germany	Raum für Kunst und Musik e.V.
2007-09-03	Düsseldorf	Germany	ARTLEIB
2007-12-07	Cologne	Germany	Raum für Kunst und Musik e.V.
2008-09-07	Düsseldorf	Germany	First Floor Contemporary
2008-10-11	Taipei	Taiwan	Aki Gallery
2010-06-26	Düsseldorf	Germany	PARKHAUS im Malkastenpark
2010-07-03	Helsingør	Denmark	Kulturhuset Toldkammeret
2010-11-13	Vancouver	Canada	304 days Gallery
2011-02-25	Munich	Germany	Tanzschule Projects
2011-03-06	The Hague	Netherlands	Nest
2011-06-23	Cologne	Germany	Philipp von Rosen Galerie
2011-07-01	Berlin	Germany	Autocenter
2011-11-18	Hamburg	Germany	salondergegenwart
2011-12-02	Cologne	Germany	Mike Potter Projects
2011-12-03	Amsterdam	Netherlands	Arti et Amicitiae
2012-01-28	Cologne	Germany	Berthold Pott
2012-02-25	Zurich	Switzerland	BolteLang
2012-03-02	Cologne	Germany	Philipp von Rosen Galerie

2012-03-03	Amsterdam	Netherlands	Amstel 41
2012-03-09	Cologne	Germany	Koelnberg Kunstverein e.V.
2012-03-22	London	United Kingdom	Rod Barton
2012-03-24	Cologne	Germany	Jagla Ausstellungsraum
2012-04-16	Cologne	Germany	Kunstgruppe
2012-04-19	Cologne	Germany	Philipp von Rosen Galerie
2012-04-26	Berlin	Germany	September
2012-04-28	Leipzig	Germany	Spinnerei
2012-07-10	New York	United States	Shoot the Lobster
2012-07-21	Düsseldorf	Germany	Philara–Sammlung zeitgenössischer Kunst
2012-10-18	Los Angeles	United States	ltd los angeles
2012-11-03	Zurich	Switzerland	BolteLang
2013-01-15	Milan	Italy	Brand New Gallery
2013-03-01	Berlin	Germany	Peres Projects
2013-03-07	Cologne	Germany	Kölnisches Stadtmuseum
2013-04-01	Brussels	Belgium	Middlemarch
2013-04-03	São Paulo	Brazil	White Cube

Note that Ostrowski, unlike the Category 1 artist, presented his art in many different galleries in many different countries. He had a "relentless and restless early search," saying yes to opportunities that came to him, no matter how far he had to travel.

Fraiberger and his team found that Category 2 artists—those who, like Ostrowski, presented in a wide range of galleries—were six times more likely to have long and successful careers.

Why was presenting to a wide range of galleries, instead of the same places repeatedly, such a predictor of success?

The researchers found in the data that there were some surprising galleries that tended to consistently give artists boosts. These places, such as the Hammer Museum, Dickinson, and White Cube, were not among the most famous galleries, and it was impossible at the time to predict which of these lesser-known galleries could give them their break. But those painters who traveled widely were likely to stumble on one of these galleries to get their break. Outsider artists who did not travel widely failed to find one of these break-making shows.

As I was learning of Fraiberger and his team's Big Data studies of artists, I was watching Bruce Springsteen's show *Springsteen on Broadway*. Springsteen described his experience at the age of twenty-one, when he had already spent many years honing his craft in rock and roll playing in bars in his hometown on the Jersey Shore. At this young age, Springsteen intuitively discovered the lesson that Fraiberger and others found in the Big Data: that talent wasn't enough and that he would have to hustle his way to be discovered. Here's how Springsteen diagnosed his problem:

> *I listen to the radio. And I think "I'm as good at that guy." "I'm better than that guy." So why not me? Answer: 'cause I live in the fuckin' boondocks . . . There is nobody here, and no one comes down here. It's a grave. . . . Who was going to come to the Jersey Shore to discover the next big thing in 1971? . . . No fucking body.*

Springsteen risked being a Category 1 artist, someone who shows his art in the same place repeatedly, hoping that someone finds them. Instead, Springsteen recognized his problem and knew the solution to make sure he was in the second category, the artists hustling all around the world trying to find the break.

In his show, Springsteen describes a meeting he held with his band at the age of twenty-one: "I gathered together the men and I said, 'We are going to have to leave the confines of the Jersey Shore and venture into parts unknown if we want to be seen or heard by anybody or discovered.' "

Through a friend who had a contact in San Francisco, he was invited to a New Year's Eve gig in three days in Big Sur, California. He and his band drove across the country in a station wagon—stopping for gas and nothing else.

Over the next few years, Springsteen lived the Category 2 artist's life: he traveled around the country, taking any gig he could get and meeting with musicians, occasionally getting an audition with a record producer who would reject him. Finally, a musician friend whom he had met through his travels got him a meeting with a musician manager he knew, which got him an audition at Columbia Records in New York City, which he nailed, leading to his first record deal. Now he was in the door, and the snowball of his career was ready to roll downhill. These days, people want to play his songs just because he is Bruce Springsteen. But when he started, nobody wanted to play his songs, because he was just some guy from the Jersey Shore.

We tend to think Bruce Springsteen is Bruce Springsteen because of his poetic lyrics and energetic concerts; we think that someone whose songs are that powerful has to be a world-famous artist. Those are necessary ingredients. But they aren't sufficient. Bruce Springsteen is also Bruce Springsteen because, at the age of twenty-one, he was willing to drive across the country for a New Year's gig to get his work seen. There are likely many musicians as talented as Springsteen who, like the Category 1 artists discovered in the Magnus data, kept performing in the same places in their hometowns, waiting and failing to be found. To be a successful artist, you need more than talent; you must be the type of person willing to drive across the country if it means you have a better chance of being discovered. Springsteen, like David Ostrowski and so many other successful artists, made his own luck.

Even if you are not an artist, the lessons of artists, as uncovered by Big Data, carry over to many other arenas.

If your field is a complete meritocracy, traveling widely to find your break may be unnecessary. The top football prospects in the world can perform on their Pro Day at their university and have all the scouts come watch them.

But many fields have more in common with art than sports. The more difficult it is to measure quality, the more what works for painters will work for you.

If you haven't yet landed that no-brainer, life-altering opportunity, you should not be like the unsuccessful artists uncovered by Big Data. You should not stay in a job with unconnected higher-ups who fail to recognize your talent. You should avoid places where talented people stagnate for decades. If it seems like your place of employment is not a Hammer Museum, a Dickinson, or a White Cube, you should bounce. If a break hasn't found you where you are yet, it is unlikely to find you there in the future.

Travel to find your lucky break!

PICASSO'S RULE: PUT MORE WORK OUT THERE TO LET LUCK FIND YOU

In a now-legendary study, Dean Simonton, distinguished professor of psychology at the University of California, Davis, found a fascinating connection. Artists who produce more work tend to have more hits. Simonton found that the relationship between quantity of art produced and greatness, measured in various ways, exists in a variety of fields.

Many of the most famous artists of all time, those with the most pieces considered masterpieces, created a shocking amount of work to get their hits.

As Adam Grant has noted in his wonderful book *Originals*, Shakespeare wrote 37 plays in a two-decade period, Beethoven wrote more than 600 pieces, and Bob Dylan has written more than 500 songs. But perhaps no artist was more prolific than Pablo Picasso. Picasso released more than 1,800 paintings and 12,000 drawings, only a tiny number of which are well-known.

Why is prolific output such a predictor of artistic success?

There are likely many reasons for this correlation. One reason: extremely talented artists may find it easier to both produce a lot of work and produce great work.

Dylan in his prime could write so many hits he sometimes forgot which ones he wrote.

One day, Dylan and his good friend Joan Baez were in a car when the radio played a song being performed by Baez. Dylan did not recognize the song, "Love Is Just a 4 Letter Word," but liked what he heard.

"That's a great song," Dylan commented.

"You wrote it," Baez responded.

Another reason for the relationship between quantity and artistic reputation: artists who have early success may have more help producing more work.

But there is an additional important reason for the connection between being prolific and success: artists who produce a lot have more chances to get lucky.

Think about it this way. The success of a particular piece of art can be a lottery ticket, with unpredictable events sometimes leading to outsized success. If you have more lottery tickets than other artists, you have more opportunities to capitalize on one of these lucky breaks.

Putting lots of work out there as an artist can be particularly important since artists are sometimes unable to predict when they have produced a masterpiece. One study that pored over letters written by Beethoven documented at least eight times in which Beethoven disliked a piece he had created, only for the world to deem it a masterpiece.

When Woody Allen finished editing his movie *Manhattan*, he was so displeased by what he saw that he asked United Artists not to release it. He even offered to make a different film for no fee to avoid the embarrassment of having *Manhattan* shown to the world. United Artists overruled Allen's verdict and put the film out there in the world, where it was immediately judged a masterpiece.

Upon completing his third album, *Born to Run*, Bruce Springsteen hated it.

"I thought it was the worst piece of garbage I'd ever heard," Springsteen has said. Springsteen considered not releasing the album and had to be convinced to release it by his producer, Jon Landau.

The album—which included the title track as well as "Thunder Road," "Jungleland," and "Tenth Avenue Freeze-Out," was a hit. It landed Springsteen on the covers of *Time* and *Newsweek,* was called by *Rolling Stone* "magnificent," and was eventually rated among the greatest rock-and-roll albums of all time.

Thankfully, Beethoven, Allen, and Springsteen put the work out in the world despite their reservations. But many artists who never reached the same level of acclaim don't do that. They pre-reject themselves.

Of course, if artists could perfectly judge which of their pieces would be well received, it would be perfectly fine for them to be exceedingly selective in what they release. But artists cannot do this. Therefore, they must avoid the temptation to limit how many times the world is exposed to their art. By putting more work out there, they can let the world sometimes surprise them by giving them a hit.

Does the value of quantity extend beyond art?

Indeed, Simonton has found a similar relationship in science—scientists who put the most papers out there are most likely to win major prizes. Scholars have found a relationship between quantity and good results in other domains as well.

PICASSO DYNAMICS IN DATING

In Chapter 1, we discussed the overwhelming evidence that certain people are more desired in dating.

You may recall one of the "no-duh" findings from that chapter: beautiful people are more likely to get responses to their messages; and people are less likely to get responses when they message beautiful people.

Here are those charts—the data on superficiality—to refresh your memory.

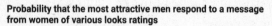

Probability that the most attractive men respond to a message from women of various looks ratings

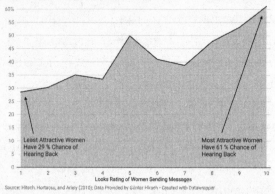

Least Attractive Women Have 29 % Chance of Hearing Back

Most Attractive Women Have 61 % Chance of Hearing Back

Looks Rating of Women Sending Messages

Source: Hitsch, Hortaçsu, and Ariely (2010); Data Provided by Günter Hirsch • Created with Datawrapper

Probability that the most attractive women respond to a message from men of various looks ratings

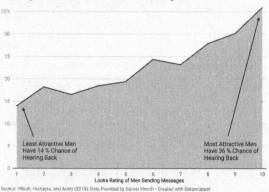

Least Attractive Men Have 14 % Chance of Hearing Back

Most Attractive Men Have 36 % Chance of Hearing Back

Looks Rating of Men Sending Messages

Source: Hitsch, Hortaçsu, and Ariely (2010); Data Provided by Günter Hirsch • Created with Datawrapper

Again, no huge surprise there. Beauty matters, the data says, in dating.

But in that chapter we didn't focus much on the actual response rates. Now let's look at those numbers closely.

Note what happens when one of the least attractive males (someone in the 1st to 10th percentile of looks) reaches out to one of the most attractive women (someone in the 91st to 100th percentile of looks).

Before seeing this data, what would you have guessed the reply rate to such men would have been? I would have guessed it would be very low. Perhaps 1 percent? Maybe 2 percent? Three percent at its very best? We are, after all, talking about men in the bottom decile of appearance asking out women in the top decile of appearance. We're talking about a 1 asking out a 10. Talk about out of your league!

In fact, men in this situation get a reply about 14 percent of the time. For women asking men out who are far more conventionally attractive, the numbers are even better. A woman in the 1st to 10th percentile of looks has about a 29 percent chance of hearing back from a man in the 91st to 100th percentile of beauty. Now, of course, not all replies lead to dates, but some of them will.

And the surprisingly-not-all-that-bad odds for people reaching well out of their proverbial league have been confirmed in other studies. Elizabeth E. Bruch and M. E. J. Newman, using different methods and data from a different dating site, found the following: When the least desirable men on the site messaged the most desirable women on the site, the response rate was roughly 15 percent. When the least desirable women messaged the most desirable men, the response rate was about 35 percent.

And these not-as-bad-as-one-might-have-thought numbers have a profound implication for the optimal dating strategy: ask out a lot of people.

Think about it this way. Suppose a man is one of the least desirable men, as ranked by Bruch and Newman, on an online dating site. He dreams of having a date with one of the most desirable women on the site. Recall from Chapter 1 that this is not necessarily a good idea for achieving long-term relationship happiness, since highly desired qualities tend not to lead to lasting relationship success. But ignore that for a moment.

He wants to date an extremely beautiful woman but knows that he is not conventionally attractive. Each time he asks such a woman out, the data tells us, he has a better chance of being turned down than being accepted.

But, since the long shot isn't as bad as he might have suspected, if he asks a number of these highly desirable people out, his chances of hearing "yes" get surprisingly high surprisingly quickly.

Here's how the math works itself out, using the estimate from Bruch and Newman that the least desirable men have a 15 percent chance of getting a reply from the most desirable women.

If a man in this situation asked one such person out, he would have a 15 percent chance of hearing back. If he asked 4 such people out, he'd have a 48 percent chance. If he asked 10 out, he'd have an 80 percent chance. And if he asked 30 out, he'd have a—wait for it—99 percent chance of hearing back from at least one of them.

Also, the numbers would be far higher for one of the least desirable women going after the most desirable men, since the data says they are even more likely to hear back. This may be one reason another study determined that heterosexual women significantly increase their odds of pairing up with a more desirable male if they initiate more contact with males.

In dating, if you take many shots, you have many chances to get lucky. Just as Picasso putting lots of art into the world allows the world to approve some of those pieces, a dater who puts more asks out there in the romantic world allows more potential partners to approve them.

And in dating, just as in art, it is important not to pre-reject yourself. You may recall the stories showing that many artists are unable to judge the quality of their own art. Beethoven thought many of his greatest pieces sucked; Woody Allen thought he would embarrass himself by releasing *Manhattan;* and Springsteen thought *Born to Run* was "garbage." The great artists put work out there anyway, despite their concerns that it won't be well received. They then get more chances for the world to give them surprisingly good news.

In dating, how many of us get stuck in our own heads, thinking that we have no chance with a man or woman we want to date? How many of us don't ask the guy or girl out because we think they are out of our league? How many of us think that we are "garbage" or that we risk embarrassing ourselves by asking certain people out?

The math says that being driven by such insecurity is clearly a mistake. The odds when you ask out people who may, on paper, seem more desirable than you may be long—but they are miles away from impossible.

I learned this data lesson in real life. In my life up until the age of thirty-five, I was far from a Pablo Picasso of asking women out. In fact, I could probably count on one hand the number of times I asked a woman out.

In graduate school, there was a beautiful and brilliant woman whom I had a crush on. I flirted with her, but I could not get myself to ask her out. It just seemed too ridiculous. Me? Her? Come on!

Many years later, I found out that she had wanted me to ask her out—and would have said yes.

Thankfully, I realized, over the years, that I had to put myself out there more if I wanted the chance of a great match.

By the time I met Julia, a beautiful, brilliant, and vivacious woman—who, more importantly, has a satisfaction with life, a growth mindset, tremendous conscientiousness, and a secure attachment style—I forced myself to ask women out even if I thought I had no chance of getting yes. I didn't allow myself to pre-reject myself. I heard "no" a lot. But I heard "yes" sometimes as well.

On Julia's and my first date—drinks on her rooftop during the Covid pandemic—I felt that she was not into me. Something in her body language told me that she wasn't feeling this. And why would she? She was two inches taller than me, more conventionally attractive than me, and more extroverted and likable.

I had my well-honed instinct to pre-reject myself. I was tempted to leave the date and never contact her again, convinced that if I asked to extend things, I would just be rejected. Instead, I overruled that instinct. Despite extreme nerves, I asked if she wanted to get dinner. Dinner turned into a second date. And a third date. And, eventually, a one-year anniversary. Sometime over the course of our relationship, I learned that she had been immediately attracted to me on that first date, and, if I had just left and ghosted, she would have been horribly disappointed and texting with all her girlfriends trying to figure out what she had done wrong. Talk about an insecure misread and dangerous desire to pre-reject oneself from *moi*!

Chris McKinlay also learned the power in dating of increasing the chances for other people to pick you and make you romantically lucky. McKinlay, whom *Wired* describes as a "math genius" who "hacked OkCupid to find true love," increased his chances to get a lucky match not by asking more people out. Instead, he came up with a clever hack to get his profile in front of more people.

McKinlay noted that women were notified anytime a man visited their profile. So, among many data-driven innovations, McKinlay wrote a bot that allowed him to visit the profiles of huge numbers of potential matches—more than he could visit manually.

Just by increasing the quantity of women who saw his profile, he could dramatically increase the number of women who were interested in him. Shortly after he began his strategy, he was getting some 400 visits per day and 20 messages per day.

This led to numerous dates, including a date with Christine Tien Wang—his 88th first date. A little over one year later, they were engaged.

Dating is a numbers game, and McKinlay hacked the system to up his numbers.

PICASSO DYNAMICS IN JOB APPLICATIONS

Even just being prolific in how many jobs you apply to can massively improve your career. A recent study surveyed hundreds of scientists about the particulars of their job search: every place they applied to, every interview they got, and every offer they received. They found that, for every offer a scientist receives, the average scientist applies to fifteen schools.

Further, the scientists who conducted the study found evidence that the scientists who were in their sample may not be applying to enough jobs. It turns out that the more jobs a scientist applied to, the more interviews they tended to receive. And scientists who received offers tended to have sent out more applications.

Think how shocking this is. Scientists work sixty-hour weeks trying to do everything possible to improve their candidacy to fulfill their dream of an academic job. But many of these same scientists do not spend the extra dozens of hours it would take to widen the pool of schools they apply to, even though evidence suggests this could improve their odds of a job.

An academic job is, to a certain degree, a lottery. And the lottery winner is likely to be the person who spends a few extra hours accumulating more lottery tickets. More quantity of applications means more chances of a quality outcome—a job.

DATA-DRIVEN DECISION MAKERS CAN DO THINGS—LIKE TRAVEL widely and put themselves out there more—that increase the odds that they are chosen for success. Or, as I like to say it, fortune favors the data-driven decision maker.

There is one more way that data-driven decision makers might up their odds that they are among the fortunate ones picked for big breaks: improve their appearance. This is such an important topic that I will devote the next chapter to it.

UP NEXT

There are new lessons in machine learning and personal data collection into how people can look their best.

CHAPTER 7

MAKEOVER: NERD EDITION

"I hate the way I look," I told my mom, at the age of six.

When I was a kid, I was teased for my appearance. My ears were too big, kids informed me. My nose was too wide, my forehead too long.

From the ages of six to thirty-eight, my feelings about my face vacillated between mild disappointment and deep despondency. In fact, during a major depression shortly after I finished my first book, *Everybody Lies*, I went to a new therapist. The first thing I told him: "I'm ugly, and this is ruining my life."

In my decades of insecurity about my appearance, I never did a whole lot to try to improve it. In fact, I responded to my unhappiness with what I looked like by putting less energy into what I looked like. I didn't take good care of my skin, dressed poorly, and went far too long between haircuts. I made a bunch of self-deprecating jokes about being ugly.

However, some months ago, I was finally convinced to take action and try to improve my appearance. In fact, I performed my own data analysis to figure out the best look for me in what may be history's nerdiest makeover attempt. You may be able to get a few tips from what I did to improve your own appearance.

The motivation for my action? A deep dive into the world of facial science.

Facial science has discovered two major things about appearance. First—and this is deeply depressing—how we look massively influences how far we advance in life. Our appearance has far more of an impact than many of us realize. Second—and this is encouraging—we can greatly improve how we look. In fact, we can improve our appearance far more than many of us realize.

YOUR APPEARANCE: IT MATTERS

Alexander Todorov, a professor at the University of Chicago, may be the world's foremost expert on faces. Todorov—who has a strong nose, mildly protruding ears, and a facial structure that appears friendly, approachable, and intellectual—researches how much a person's facial appearance influences that person's success in a variety of domains. (Hint: a lot.)*

Consider, for example, that all-important domain of politics. We would like to think that the winners of major political elections might be somehow the most deserving. They do, after all, determine how trillions of dollars are allocated. We might hope that the men and women whom we elect to make

---

* *Todorov has written a fascinating book called* Face Value, *which I highly recommend.*

these decisions have the greatest intellect. Perhaps our political leaders work the hardest or have come up with the wisest policy positions.

Alexander Todorov and others have proven, however, that the winners of major elections frequently merely impress voters with their face.

In one study, Todorov and colleagues collected photos of the faces of the Democratic and Republican candidates in a large sample of Senate and House races. They recruited a group of subjects to simply say which of the two candidates in each race looked more competent. (They did not include subjects if they recognized the faces.)

If you like playing along with scientific studies or enjoy ranking people's faces, you can answer one of the questions the subjects of the researchers' study were asked.

Of the two people shown below, who do you think looks more competent?

All politician photos are from FiscalNote/Congress at Your Fingertips. They are used with permission.

Got your answer?

I am guessing that you picked the gentleman on the right as Monsieur Competent.

If you indeed thought that the person on the right looks more competent, you are in agreement with the vast majority of people to whom Todorov and colleagues showed these pictures. Ninety percent of subjects said that the man on the right looked more competent than the man on the left. Further, it did not take them particularly long to come to that conclusion. The average person judged the person on the right as more competent in roughly one second.

So, who are those two men?

These were the two candidates for the 2002 United States Senate election in Montana. The man on the right, the one who looks more competent, is Max Baucus, the Democratic candidate in that election. The man on the left is Mike Taylor, the Republican candidate.

Baucus, the man who 90 percent thought looked more competent, won the election with a resounding 66 percent of the two-party vote. In other words, the candidate whom people judged, upon viewing them for all of one second, as looking more competent was elected by voters.

This—the candidate who looks more competent also being the winner of the election—is the start of a pattern.

Ready to play along some more with the researchers' experiment? Look at the two pairs of faces on the following page. For each, determine which of the two people next to each other strikes you as more competent.

Once again, I have a guess for your answers. I guess that,

VS.

VS.

among the two gentlemen on top, you chose the man on the right as looking more competent. Among the lady and gentleman on the bottom, I guess you chose the man on the left as looking more competent.

If those were your choices, you agree with most people.

They were the choices of roughly 90 percent of people shown these pairs of faces.

In other words, the faces that most people say look more competent won their elections. The man on the top right is Republican Pat Roberts. He won the 2002 Kansas race with 82.5 percent of the vote, defeating the man on the left, libertarian Steven Rosile. The man on the bottom left is Republican Judd Gregg; he defeated the woman on the bottom right, Democrat Doris Haddock, in the 2004 New Hampshire Senate race, winning 66 percent of the vote.

In fact, across all the races that Todorov and colleagues studied, they found that the person whose face was judged as more competent by the majority of subjects won 71.6 percent of the Senate races and 66.8 percent of the House races. And the importance of looking competent for winning elections held even taking into account other factors, such as ethnicity, age, and gender.

Some people, the work of Todorov and others tell us, have a face that screams competence. Others don't. Voters tend to choose the candidate who looks the part. Or as Todorov and coauthors summed up the results of their study: "[Voters are] more shallow than we would like to believe."

James Carville, the campaign strategist for Bill Clinton in 1992, while discussing what voters most care about, famously said, "It's the economy, stupid." But research tells us, when it comes to winning over the public for major elections: it's the face, silly.

POLITICS ISN'T THE ONLY ARENA IN WHICH ONE'S FACE CAN DEtermine how far you rise. Nor is competence the only judgment we make about people from their face. In looking at people, we judge how trustworthy they are; how smart they are; how extroverted they are; how energetic they are; and much more.

In politics, Todorov and others have found, it is the appearance of competence in a face that counts the most. But, in other domains, other aspects of one's face—and other attributes we assign to them based on these aspects—play a more important role.

Consider, for example, the military.

Researchers were interested in seeing what best predicted which West Point cadets had the most career success. The researchers created a database that included the rank each cadet had reached twenty years after graduation along with various attributes of the cadets in school.

They collected data on how wealthy the cadets were when they were growing up, how they performed on various academic tests in schools, and how they performed in various athletic endeavors. Finally, the researchers collected the graduation picture of each cadet and asked people to rate how they came across.

The researchers found that one fact above all predicted how far a cadet rose in his career. It wasn't how distinguished their family was, how smart they were, or how fast they ran. In fact, each of these traits only weakly predicted cadets' career performance.

The biggest predictor of cadets' career success was how dominant their faces appeared. Having a face that people judged as looking dominant increased the odds of a colonel becoming a brigadier general, a brigadier general becoming a major general, and a major general becoming a lieutenant general.

A Face of Dominance   *A Face Lacking Dominance*

Among people good enough to get into West Point, in other words, those who look dominant tend to be allowed to dominate.

THE STUDIES—AND THERE ARE MANY OF THEM—SHOWING THE relationship between our face and how we are treated—are, I admit, depressing. It is sad that human beings can be so superficial and that our superficiality can have such large consequences.

These results also seem hopeless for those of us who may not have been blessed with the best face. If we have faces that don't make people think we are competent, is a career in politics off the table? If we have faces that appear less dominant, are our military dreams impossible?

Not exactly.

There is an interesting twist to this research. And, like so many twists to scientific literatures, this science twist was best encapsulated in an episode of *Seinfeld*.

YOUR APPEARANCE: IT VARIES

In the tenth episode of season nine of *Seinfeld*, Jerry is dating a woman, Gwen, who is a "two face." Sometimes she is very attractive. Other times she is very unattractive. For Gwen, lighting seems to play the key role in whether she is an 8 or a 2.

In the middle of the episode, Jerry introduces Gwen to his friend Kramer, during a moment in which she looks unat- in the episode, Kramer runs into a woman on m he does not recognize and finds very attrac- an, who is actually Gwen in good lighting, tells he is dating Jerry. Kramer explains to her that bly be true. He explains that he has recently met d, who is much less attractive than the woman ing to him.

Gwen now becomes convinced that Jerry must have another girlfriend and is cheating on her with this less attractive woman whom Kramer has described.

In one of the final scenes of the episode, Gwen, again looking attractive, confronts Jerry. She tells him that she knows that he is cheating on her with a less attractive woman and storms out of the room. Jerry chases her down to explain that he is not cheating and woo her back. However, when he reaches Gwen, on a porch, she now looks unattractive, killing Jerry's desire to be with her. Jerry turns around and walks away.

"Bad lighting on the porch," Jerry explains.

Science tells us that we are all, to some degree, Gwen, sometimes looking better and sometimes looking worse.

IN ALL THE PREVIOUS STUDIES THAT I HAVE DISCUSSED, TODOROV and other researchers in the world of facial science have been asking people to rate one particular photo of a person. One photo of Judd Gregg was compared to one photo of Doris Haddock. People determined, from the one photo of each, whether Gregg or Haddock looked more competent. One photo of each person from the military was studied. People determined, from these photos, how dominant each cadet looked.

It was as if each person had a set look of competence or dominance or attractiveness—and there was nothing they could do about it.

But in a fascinating study with Jenny M. Porter of Columbia University, Todorov asked people not to rate just one picture of a person. The researchers asked people to rate multiple pictures of the same person on many dimensions, including competence, attractiveness, and trustworthiness.

They used a dataset that was created to help with facial recognition and included five to eleven head shots of the same person. The pictures of each individual were only slightly different.

Despite the similarity of the pictures of the people, the particular picture shown led to great variation in how people perceived them. For example, the pictures that follow show two different pictures of two men. People were asked to rank the trustworthiness of the face in each photo. The man who was ranked most trustworthy switched depending on the particular photo shown.

Trustworthiness

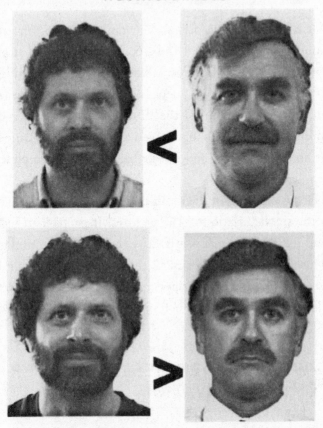

Pictures are from the FERET database and used with permission. They originally appeared in P. J. Phillips, Hyeonjoon Moon, S. A. Rizvi and P. J. Rauss, "The FERET evaluation methodology for face-recognition algorithms," in *IEEE Transactions on Pattern Analysis and Machine Intelligence*, vol. 22, no. 10, pp. 1090–1104, Oct. 2000, doi: 10.1109/34.879790.

This pattern repeated. How people were perceived could vary depending on what picture of them others were shown.

For attractiveness, a person who ranked as a 5, on average, might vary between a 4 and a 6, depending on what picture was shown. A person who ranked a 3, on average, might vary between a 2 and a 4, depending on what picture was shown. For other qualities, the differences were even more pronounced. Someone who averaged a 5.5 on trustworthiness might rate between a 4 and a 7, depending on which picture of them was shown.

The big variation in the ratings was even more remarkable given the small variation in the pictures. If people can move between a 2 and a 4 in how attractive they are just from slight differences in lighting and smiles, this implies there might be even more variation from bigger changes: changes in facial hair, haircuts, glasses, and much more.

THE WORLD'S NERDIEST MAKEOVER ATTEMPT: MOTIVATION

Can I use data to improve my appearance? That was my first thought upon reading the research in facial science. Improving my appearance is somewhat of a revolutionary thought for me. As mentioned, I have always considered myself unattractive and assumed that was a fixed trait of mine. But Todorov's and Porter's work shows that different versions of faces can be perceived very differently. Perhaps I could find the version of my face that is perceived best by the world, I wondered.

But how should I find this version of my face?

Well, I didn't want to just rely on my gut. Decades of research in psychology has shown that people are not particularly accurate judges of how they come across. There are numerous biases that get in the way of seeing ourselves clearly. And if anybody is going to be bad at seeing how they come across physically, it is going to be me. Clearly, I would need some outside opinions.

The solution I stumbled on to try to improve my appearance used three (very modern) steps: artificial intelligence, rapid market research, and statistical analysis. What I'm trying to say is, while I may not have attached ears, a small nose, or a standard forehead, I sure as hell do have the ability to do statistical analysis to improve my appearance!

THE THREE-STEP PLAN TO LEARN WHAT MAKES YOU MOST ATTRACTIVE WITHOUT LOGGING OFF YOUR COMPUTER

Step 1 (Artificial Intelligence): I downloaded FaceApp, an app that uses artificial intelligence to alter a picture. If you don't know FaceApp, here's how it works. You upload a picture and then can change the settings to alter the picture in shockingly realistic ways. You can adjust your hairstyle, hair color, facial hair, glasses, and smile.

Different (AI-Generated) Versions of Me

I created more than one hundred different versions of my face. Above are some of the looks that I ended up with.

Step 2 (Rapid Market Research): I conducted rapid market research on different versions of my face. For this, I used GuidedTrack and Positly, two programs created by my friend Spencer Greenberg to allow anybody to quickly and cheaply perform survey research. For each picture, I asked people to judge how competent the person in the picture looked on a scale of 1 to 10. (You can also use the site Photofeeler.com to rate different photos.)

I discovered that there were massive differences in how people ranked me in the different FaceApp photos. For example, in the picture directly to the right, I ranked 5.8/10 in competence, among the lowest scores of those I tested.

In the picture directly below it, I ranked 7.8/10 in competence, the highest score that I received.

Just as Todorov and Porter had discovered and just as *Seinfeld* had pointed out, I can be perceived very differently.

Step 3 (Statistical Analysis): I used R, a statistical programming language, to recognize patterns in how different styling decisions I make affect how I am perceived. This allowed me to figure out which aspects of me have the most impact on how I am perceived.

So, what did I learn?

The biggest boost I receive in perceived competence comes from wear-

ing glasses. I gain, on average, about 0.8 points on a 10-point scale when I wear glasses. This was surprising to me, as I had concluded that I look terrible with glasses and was moving toward wearing contact lenses as much as possible. The data, in other words, told me that I should overrule my instinct to avoid glasses.

The next biggest competence boost I get is from having a beard. I gain, on average, about 0.35 points of perceived competence with a beard. I had never grown a beard in my life until the age of about thirty. Over the past five years, I have gone back and forth between beard and no beard. But the data is clear that, for me, beards are a positive.

Besides glasses and beards, the other changes didn't make a huge difference. Slight differences in hairstyle and different hair colors did not have a statistically significant effect on how I was perceived. The one, perhaps obvious exception is that pink hair cost me about 0.37 points in perceived competence.

I had been worried that I should smile more in pictures or find a better smile. But there was no statistically significant difference in my perceived confidence from smiling. I found this comforting and am going to worry less about my smile in pictures.

When it comes to how I am perceived, glasses and beards are the game changers—and, as long as I keep my hair non-pink, nothing else I do matters all that much.

Thanks to AI plus rapid market research plus statistical analysis, I feel that I have my look. I will wear glasses and a beard. That version of me might not be able to compete with Mitt Romney or Barack Obama at the highest levels of American politics. But that version of me is certainly, the data suggests, a person who could give a good first impression. And the thought that a version of me could score a 7.8/10 on any 10-point scale related to appearance would have been inconceivable to twenty-five-year-old me.

So, what can you take away from my nerdy makeover?

I realize that I took this project to a somewhat extreme level—using a spreadsheet and statistical analysis. But I believe that just about anyone could benefit from a less extreme version of this analysis.

At a minimum, you can download FaceApp or a similar program and examine many different looks for you. Even if you don't utilize a random sample of people on the internet, you can ask people on social media or your more honest friends to pick which look is better on you.

I am convinced that this type of analysis is far better than the (flawed, gut-driven) approach that most of us use to learn about our appearance.

Often, we fixate on problems that are entirely in our head—such as my concern about my smile. Many of us go through large parts of our lives never even thinking of a style that might dramatically improve our appearance—such as my decade living life without a beard. And most of us can't see ourselves accurately—such as my being convinced that glasses look horrible on me.

What I'm trying to say is, the research is overwhelming that our appearance matters, that we can improve our appearance, and that we are poor judges of our own appearance. Or, in other words, AI plus rapid market research plus statistical analysis dominates mirrors.

UP NEXT

The past four chapters have focused on how to be more successful in your career. If you use this advice, you may find yourself more successful. But you may also find yourself, like so many successful people, absolutely miserable. Thankfully, new data can also give people advice on how to be happier.

CHAPTER 8

THE LIFE-CHANGING MAGIC OF LEAVING YOUR COUCH

What makes human beings happy? Some of the greatest philosophers in history have tried—and largely failed—to answer this question. Could the answer finally be found in . . . our iPhones?

No, the answer to happiness is not to (over) use iPhones, which most definitely make people miserable (more on that later). Instead, the answer to happiness may come from the research that has become possible thanks to iPhones and other smartphones.

Researchers such as George MacKerron and Susana Mourato, who have led the Mappiness project, which I briefly mentioned in the Introduction, have recruited tens of thousands of smartphone users to help in the quest to understand happiness. Mappiness has pinged their users at random times during the day to ask them simple questions about what they are doing and what their mood is.

From these simple questions, the researchers have collected more than 3 million happiness points—many times more than previous happiness datasets.

So, what do 3 million–plus happiness points teach us about what makes people happy? I will answer that question shortly. But first I will discuss the state of happiness research prior to smartphone-based projects such as Mappiness. One of the major lessons from the small, survey-based studies that were the norm prior to smartphones is that we human beings stink in understanding what makes us happy—and were in dire need of projects such as Mappiness to give us some happiness pointers.

MISPERCEPTIONS OF WHAT WOULD MAKE US HAPPY

How happy do you think you would be if you got your dream job? How miserable would you be if your preferred political candidate lost? How about if a romantic partner dumped you?

If you are anything like the average person, your answers to these questions are something along the lines of "very happy if I got my dream job"; "very miserable if my candidate lost"; and "even more miserable from being dumped."

Your answer to every one of these questions is likely to be wrong. So reported a groundbreaking study in happiness research by Daniel Gilbert and colleagues.

There were two parts to the study.

In the first part, the researchers asked a group of people questions like those that I posed to you in the first paragraph. For example, in one experiment the researchers recruited a group of assistant professors who were all hustling to get their dream job: a tenured professorship. The researchers asked how much their future happiness depended on the tenure decision. In particular, they were asked to imagine changes in their happiness under the two possible life paths. Life Path 1: they got tenure. Life Path 2: they were denied.

As someone who has spent much of my adult life around assistant professors who do little else but eat, sleep, and try to get tenure, I found the results here not at all surprising. The assistant profs estimated that they would be substantially happier under Life Path 1 than Life Path 2. Getting tenure would lead to many happy years, assistant profs said.

In the second part of the study, the researchers cleverly recruited a different group of subjects: people from the exact same university who earlier had been up for a tenure vote. These were people who had gone down the different life paths that the first group of assistant profs was approaching. Some of these people had gotten the big prize (tenure). Some of them had not.

The researchers asked all these people to report how happy they were now. The results? There was no significant difference in the reported happiness of those who received tenure and those who had been denied.

In other words, the data from the pre-tenure subjects proved that academics think that tenure will give them a years-long boost in happiness. The data from the post-tenure subjects proved that tenure does not give that boost.

Elliot Ferguson* lived this lesson. He recently answered a question on Quora asking what it is like being denied tenure. He said that he was "devastated" in 1976 when he was denied tenure in psychology from the University of Wisconsin-Madison. Having fully devoted himself to getting tenure, he had not prepared himself at all for this outcome. But he, like many human beings, proved resilient. He built a career in business, as an entrepreneur and a consultant. He enjoyed working with the "bright, creative, and interesting people outside of academia" and appreciated the ability of businesses to get things done. Thirty-seven years after the tenure denial, here's what he has to say: "So, I say, thanks University of Wisconsin for denying me tenure. It was the right thing to do for yourself and the best thing to do for me."

The data from Gilbert and his coauthors suggests that the Ferguson story is representative. Academics bounce back from not getting tenure, even if they think that they won't.

It isn't only academics who are looking to climb the career ladder who fail to predict how they would respond to life events. Gilbert and the other researchers used the same method to test whether people could predict how happy they would be in the aftermath of other major events: romantic heartbreak and political developments, for example.

People consistently predicted that these events would lead to major changes in their happiness. But people who had experienced these events reported that it made no major difference

---

* *His name and some details have been changed.*

in their long-term happiness. In other words, events that seem to us like they would be awful and irrevocably bad frequently turn out to be not such a big deal.

So why are we so bad at predicting what makes us happy? Part of the problem is that we are really bad at remembering what made us happy—or miserable—in the past. It is difficult, after all, to predict what we can't remember. How do we know that we are bad at remembering our past feelings? Evidence from this comes from an important study that is both extremely clever and extremely stomach-turning.

MISPERCEPTIONS OF WHAT MADE US HAPPY

Strange quiz. Two people—call them Patient A and Patient B—are both getting a colonoscopy. During the colonoscopy, they are asked to record, every sixty seconds, just how much pain they are in—on a scale of 0 to 10. (This is called moment utility.) At minute zero, they are asked, 0 to 10, how bad is it? At minute 1, they are asked the same. And so on. Until the colonoscopy is done.

When the colonoscopy is finished, we now have a pain chart for both patients, which allows us to see how much pain each was in for every minute of their colonoscopy. The pain charts are shown on the following page.

Patient A, as you can see in their chart, had a pain level that fluctuated between 0 and 8 for roughly 8 minutes. Patient B, as you can see in their chart, had a pain level that fluctuated between 0 and 8 for more than 20 minutes.

Patient A's pain intensity

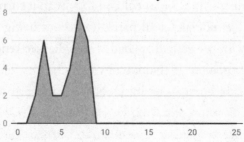

Source: Redelmeier and Kahneman (1996) • Created with Datawrapper

Patient B's pain intensity

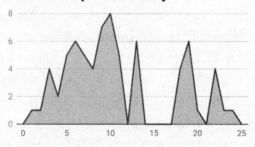

Source: Redelmeier and Kahneman (1996) • Created with Datawrapper

Now, the strange quiz question: who experienced more total pain during their colonoscopy, Patient A or Patient B?

See the charts? Got your answer?

This isn't a trick question. The answer is obvious: Patient B experienced more pain. Patient B was in roughly as much pain as Patient A for the first 8 minutes and then experienced 17 minutes of additional pain on top of all that. By any metric, Patient B had the more painful colonoscopy. If you chose Patient B, you scored an A+ on this strange quiz. Good job!

Why am I asking you this obvious, easy question?

Because while it may be easy for us, while looking at actual data that the patients gave us during their colonoscopies, to say how bad a particular colonoscopy was, it turns out to be very difficult for the actual patient, without being shown the data, to recall precisely how bad it was. People tend to forget just how painful their colonoscopy was.

The evidence: a paper by Donald Redelmeier and Daniel Kahneman where these charts were shown.

The researchers recruited a whole bunch of colonoscopy patients and asked them to record their pain for every minute of the procedure, producing moment utility charts like those shown above.

But what really made this paper special was another question that the scholars asked. They asked every patient, well after the procedure had ended, just how bad the experience was. The patients were asked both to rate the pain on a numerical scale and to compare it to other bad experiences in their lives. This is called remembered utility.

Here is where things got interesting.

Take Patient A and Patient B. Recall that Patient B's moment utility chart clearly showed that they experienced more pain than Patient A. But after the fact, Patient B recalled that they had experienced less pain than Patient A. In other words, the patient who experienced more pain, for more time, recalled, incorrectly, that they had experienced less overall pain.

Also, the disconnect between moment utility and remembered utility wasn't limited to these two patients. Redelmeier and Kahneman found that there was only a small relationship between how horrible the colonoscopy was in the moment and how horrible the patients remembered the colonoscopy to be. Quite simply, many people who had experienced less pain recalled afterward that they had experienced more pain (and vice versa).

COGNITIVE BIASES THAT DERAIL OUR MEMORIES OF PLEASURE AND PAIN

Why do human beings tend to be bad at remembering how bad (or good) an experience was? Scientists have discovered many cognitive biases that we *Homo sapiens* have that interfere with our ability to correctly remember the pleasures and pains of our past.

One major mental bias that interferes with our happiness recall: duration neglect. This bias means that, in judging the quality of a past experience, we fail to take into account how long that experience lasted. Clearly, in real time, people want pleasurable experiences to last longer and painful experiences to end quicker. Clearly, in real time, a person would want, say, a painful colonoscopy to be as short as possible. But, after the fact, duration neglect tells us, people tend not to distinguish between painful experiences of differing lengths. People just remember that an experience was bad but don't remember how long it was bad for. Five minutes of pain is hard to distinguish, in hindsight, from fifty minutes of pain.

Duration neglect explains part of the reason that Patient B failed to remember the unusually painful nature of their colonoscopy. One of the reasons that it was so bad was that it lasted so long.

In fact, in Redelmeier and Kahneman's study, there was almost no relationship between how many minutes the colonoscopy lasted and how painful people remembered it being. Some people experienced colonoscopies of a mere four minutes; others experienced colonoscopies that lasted more than an hour. But after the fact, it was all a painful colonoscopy to everybody.

Duration neglect, interestingly, can make it difficult to test the effectiveness of medications. If a particular medicine decreases the duration of, say, a patient's migraines from twenty minutes to five minutes, this would be an enormous effect. But a patient may not notice it—and may not report to a doctor an improvement. Due to duration neglect, many health scholars recommend patients carefully record the length of their symptoms before and after an intervention to see if they might be improving without the patients' realizing it.

Another cognitive bias that tricks us from making sense of our past experiences is the peak-end rule. We tend to judge past experiences not based on their overall pleasures and pains over the entire experience. Instead, we give undue weight to the peaks of the experience (how high the highs were or how low the lows were) and the ends of the experience (whether it ended on a high or low note).

Return to the moment utility charts of Patients A and B. You can see that, while Patient B's colonoscopy led to more pain over a longer period, the second half of the colonoscopy was less painful than the first half. This tricks Patient B into underestimating how awful the colonoscopy was.

In fact, Redelmeier and Kahneman found that a key factor in predicting how painful people remembered their colonoscopy as being is how much pain they experienced in the final three minutes of the procedure.

Since we suffer from duration neglect, the peak-end rule, and other cognitive biases, it is no wonder that human beings are not so great at learning from our own experiences to study our own happiness.

The same problems that have hindered an individual's attempt to figure out what makes them happy have, historically, hindered scientists' attempts to understand happiness. Scientists have usually only been able to interview a small number of people a small number of times. Frequently, scientists have attempted to ask people to report how happy they were performing various tasks. But, as mentioned, people can misremember how happy they were.

Redelmeier and Kahneman were able to learn about patients' experiences of colonoscopies by asking 154 people to record their moment utility for the minutes on the day they got a colonoscopy. The ideal study of happiness would instead ask a large number of people to record their moment utility during many days, in which they engaged in many different activities.

That wasn't possible for most of human history. It was possible once iPhones were invented.

THE IPHONE: A REVOLUTIONARY TOOL THAT ALLOWED FOR CONVINCING RESEARCH SHOWING HOW MISERABLE IPHONES MAKE PEOPLE

Some years ago, George MacKerron, a senior lecturer in economics at the University of Sussex, and Susana Mourato, a professor of environmental economics at the London School of Economics, had an insight. Thanks to people carrying around smartphones, researchers may be able to dramatically scale the availability of moment utility charts. Instead of recruiting users to fill out a pen-and-pencil survey of how they are feeling, they could just ping them on an app.

They created an app, Mappiness, recruited users, and

pinged these users at different points of the day to ask them some simple questions. The questions included:

» What are you doing? (Users could select from forty activities, everything from "Shopping/errands" to "Reading" to "Smoking" to "Cooking/Preparing Food.")
» Who are you with?
» How happy do you feel, on a scale of 1 to 100?

So, did this project succeed in bringing happiness research into the age of Big Data?

You betcha. After some years, the Mappiness team had built a dataset that contained more than 3 million happiness measures from more than 60,000 people. This was like the moment utility charts that Kahneman, Redelmeier, and others had pioneered—but on steroids.

MacKerron, Mourato, and coauthors have done all kinds of fascinating studies that can only be done with this kind of rich data; some of their more interesting studies joined together the Mappiness data with external datasets on things such as the weather or the environment. Some of these studies will be the focus of the next chapter.

But this chapter will focus on the Mappiness basics: a study of the happiness of forty activities. Recall that Mappiness asks users both what they are doing and how much happiness they are experiencing. This—combined with the enormous sample size—allowed MacKerron and a coauthor, Alex Bryson, to estimate how much each of the forty activities tends to contribute to happiness. They produced what I call the Happiness Activity Chart, which I believe must be frequently consulted by any data-inclined person deciding how to spend their time.

Crucially—and this is a bit technical—Bryson and MacKerron didn't merely average the happiness levels of every activity for every person doing them. Instead, they used statistical techniques that allowed them to compare the same person, at the same time of day, doing different activities. This made it more convincing that they were estimating the causal role of the activity on happiness, not merely documenting a correlation.

Okay, it's time to reveal what this revolutionary study of the happiness of various activities reveals! Let's start with the activity that gives people the most happiness. Ready to learn the activity that gives people the most happiness? The activity that gives human beings the most happiness is . . .

. . .

(drum roll)

. . .

(pausing to build suspense)

. . .

(still pausing to build suspense)

. . .

Okay. Yes, it's sex.

People who reported to Mappiness that they were making love were happier than any other group of people, easily besting people engaged in the second-place activity, attending a show.

Now, initially sex's place on the top rung of the Happiness Activity Chart doesn't seem all that surprising. Of course, sex makes people happy. Natural selection, after all, did the best it could to make sex as pleasurable as possible. Further, somewhere in the back of my mind, I hear the cool kids from high school saying, "While you nerds were spending years applying for grants, designing questionnaires, and coding your apps to uncover that sex is enormously pleasurable, we were busy, ya know, having sex." Touché.

But, when you pause to think about Mappiness's methodology, the popularity of sex in the dataset is actually surprising. Recall that Mappiness only collects data on people willing to answer their survey the moment they hear a ping.* There is what statisticians call a selection bias here: the only sex participants in the Mappiness data sample were those willing to stop and answer the question.

It is safe to assume that people in the midst of heart-pounding, mind-blowing, furniture-shaking, floor-rattling, scream-inducing, neighbor-awakening sex were able to tune out the little ping from Mappiness. Mappiness's sample of sexual participants instead consists of people having sex so underwhelming that they are willing to stop in the middle, pick up their phones, and answer a series of survey questions. And even these people—sex's most indifferent partakers—were happier than any group of people trying any other activity. Bad sex literally beats anything else human beings can think to do.

Thus, lesson number one of the data science of happiness: *have more sex, people!!!* Even, it seems, if you're looking at a phone during it.

---

* *This is an exaggeration for attempted comic relief. Mappiness does allow users to respond for up to an hour after the ping and say what they were doing while they were pinged and how happy they were.*

After learning of this data-driven lesson, I got excited and told my girlfriend that we need to show this finding to my good friend. My friend's girlfriend has been complaining recently that he never wants to have sex. He often claims that he is too tired or needs to work. But maybe, I told my girlfriend, if he were just informed of the data, he would stop making excuses and better satisfy his partner. My girlfriend looked me in the eye, scowled, and said, "We need to show this finding to *you*." That is all I'm going to say in this book about my sex life—or my inability to satisfy a woman.

Fine, fine, fine. I'll say a bit more about my inability to satisfy a woman. That night, my girlfriend reminded me of the finding. We had sex. For a few minutes. She stopped in the middle to answer a Mappiness survey.

Anyway, enough about sex. What else did Mappiness tell us?

THE FULL LIST OF HAPPINESS-PRODUCING ACTIVITIES

Here are the rest of the results on how happy various activities make people—all from the Mappiness data analyzed by Bryson and MacKerron. Afterward, we will discuss the implications of the findings.

HAPPINESS ACTIVITY CHART

Activity Rank	Activity	Gain in Happiness Relative to Not Doing Activity
1.	Intimacy/Making Love	14.2
2.	Theater/Dance/Concert	9.29
3.	Exhibition/Museum/Library	8.77
4.	Sports/Running/Exercise	8.12
5.	Gardening	7.83
6.	Singing/Performing	6.95
7.	Talking/Chatting/Socializing	6.38
8.	Birdwatching/Nature Watching	6.28
9.	Walking/Hiking	6.18
10.	Hunting/Fishing	5.82
11.	Drinking Alcohol	5.73
12.	Hobbies/Arts/Crafts	5.53
13.	Meditating/Religious Activities	4.95
14.	Match/Sporting Event	4.39
15.	Childcare/Playing with Children	4.1
16.	Pet Care/Playing with Pets	3.63
17.	Listening to Music	3.56
18.	Other Games/Puzzles	3.07
19.	Shopping/Errands	2.74
20.	Gambling/Betting	2.62
21.	Watching TV/Film	2.55
22.	Computer Games/iPhone Games	2.39
23.	Eating/Snacking	2.38
24.	Cooking/Preparing Food	2.14
25.	Drinking Tea/Coffee	1.83
26.	Reading	1.47
27.	Listing to Speech/Podcast	1.41
28.	Washing/Dressing/Grooming	1.18
29.	Sleeping/Resting/Relaxing	1.08
30.	Smoking	0.69
31.	Browsing the Internet	0.59
32.	Texting/Email/Social media	0.56
33.	Housework/Chores/DIY	−0.65
34.	Traveling/Commuting	−1.47
35.	In a Meeting, Seminar, Class	−1.5
36.	Admin/Finances/Organizing	−2.45
37.	Waiting/Queueing	−3.51
38.	Care or Help for Adults	−4.3
39.	Working/Studying	−5.43
40.	Sick in Bed	−20.4

Source: Bryson and MacKerron (2017)

All right. So, what should you do with this list?

Well, if you are as nerdy as I am—and I fear that nobody is as nerdy as I am—you may want to take a photo of this chart, upload it on collage. com or a similar service, and order an iPhone case with the Happiness Activity Chart printed on it.

	...ove	14.2
	...ncert	9.29
	...eum/Library	8.77
4.	Sports/Running/Exercise	8.12
5.	Gardening/Allotment	7.83
6.	Singing/Performing	6.95
7.	Talking/Chatting/Socializing	6.38
8.	Birdwatching/Nature watching	6.28
9.	Hunting/Fishing	5.82
10.	Drinking Alcohol	5.73
11.	Hobbies/Arts/Crafts	5.53
12.	Meditating/Religious Activities	4.95
13.	Match/Sporting Event	4.39
14.	Childcare/Playing with children	4.1
15.	Pet care/Playing with pets	3.63
16.	Other games/Puzzles	3.07

Now, every time I am considering an activity, I can look at the back of my phone, see how much happiness I can expect from said activity, and make a data-driven decision as to whether to partake.

Back to the actual chart—and how to interpret it. Of course, some of the results on how pleasurable different activities are obvious. You probably didn't need scientists to tell you that having an orgasm is more pleasurable than having the flu.

However, some of these findings might not have been so obvious before the Mappiness project. Before reading the chart, would you have known that watching TV makes people so much less happy than gardening? Would you have known that relaxing delivers so much less pleasure than birdwatching? Would you have known that cooking tends to be worse for one's happiness than arts and crafts? Most people, it turns out, did not know these facts.

UNDERRATED AND OVERRATED ACTIVITIES

Another social scientist, Spencer Greenberg, the founder of clearerthinking.org, and I were curious if people could accurately guess the ordering of the activities in the Happiness Activity Chart. We asked a sample of people to guess how happy people were, on average, doing each of the activities that MacKerron and Bryson studied.

What was the motivation for our study? One idea we had is that, if people systematically overestimate the happiness that an activity offers, you should probably have a bit more skepticism in how often you engage in that activity. If people think that an activity makes people happier than it does, then you may be subject to the same bias and should be a little more wary in engaging in that activity. On the flip side, if people systematically underestimate the happiness that an activity offers, you should probably have a bit more enthusiasm to engage in that activity. In other words, a wise life hack is to do things that tend to make people happier than they realize.

So, what were the results? Were people able to predict how happy different activities tend to make people?

On balance, people got many of the activities right. Again, the Happiness Activity Chart isn't so shocking. People correctly realized that sex and socializing were near the top in giving happiness boosts and that being sick in bed and working were near the bottom.

But there were some activities for which people misjudged how much happiness they tend to give people. Here are the biggest misjudged activities:

Underrated Activities: These Tend to Make People Happier Than We Think[*]

» Exhibition/Museum/Library
» Sports/Running/Exercise
» Drinking Alcohol
» Gardening
» Shopping/Errands

Overrated Activities: These Tend to Make People Less Happy Than We Think

» Sleeping/Resting/Relaxing
» Computer Games/iPhone Games
» Watching TV/Film
» Eating/Snacking
» Browsing the Internet

So, what should we make of those two lists? "Drinking alcohol" is obviously a complicated route to happiness, due to its addictive nature; I will talk more about the relationship between alcohol and happiness in the next chapter.

But one systematic bias people have is they seem to overestimate the happiness effect of many passive activities. Think of the activities on the "Overrated Activities" list. Sleeping. Relaxing. Playing games. Watching TV. Snacking. Browsing the internet. These are not exactly activities that require a lot of energy.

Our minds seem to trick us into thinking that such passive activities give us more pleasure than they actually do. Ask people, as Greenberg and I did, how happy people are when

---

[*] *The full results can be found in the Appendix of this book.*

they are doing these passive activities. And ask people, as Mappiness did, how happy people are when they are doing these passive activities. There is a disconnect. People suspect these activities cause more happiness than they do.

On the flip side, a lot of activities on the "Underrated Activities" list require some energy to get started. Going to a museum. Playing sports. Exercising. Going shopping. Gardening. These require getting off your couch. And some of the activities that require getting off your couch feel like they aren't going to make us as happy as they actually do.

In fact, the study with Greenberg forced me to do something that I really hate to do: disagree with Larry David.

I HATE TO DO THIS, BUT . . . WHY LARRY DAVID IS PRETTY, PRETTY WRONG

I was once watching the comedian Larry David speak on a YouTube video when he humorously spoke of a feeling I—and perhaps you—could relate to well: the joy that comes when plans are canceled. As David put it, "If somebody cancels on me, that is a celebration. . . . You don't have to make up an excuse. It doesn't matter! Just say you're canceling. I'll go 'Fantastic.' I'm staying home. I'm watching TV. Thank you!"

Now, I am a huge Larry David fan. Some people's life motto is "What Would Jesus Do?" My life motto is "What Would Larry Do?" So, yeah, I am a huge Larry David fan. But the thesis of this book is that even great minds, absent data, make incorrect judgments. David, as brilliant and humorous as he may be, is no different. We can't trust anybody's gut, even Larry David's. And David seems to be falling for a trap that many of us do: to exaggerate the value of not doing things.

The Mappiness data makes clear that many passive activities, such as watching TV, don't yield much happiness—and lead to less happiness than people expect.

One of the best ways to improve one's happiness is to avoid that instinct to avoid doing things that seem like a lot of energy. When the thought of doing an activity makes you go "ughhh," that is likely a sign you should do it, not that you shouldn't.

When someone used to cancel a plan to go to a show together or have a dinner party together or go for a run together, I used to say, "What would Larry do?," thank my blessings for the cancellation, and surf the internet by myself. Now instead I say, "What would the Mappiness data say?" And I look at my iPhone case and try to overrule my instinct to sit on my couch and passively consume media. Mappiness data tells us there is great value (and more value than most suspect) in leaving your couch—unless, of course, you are having sex on that couch.

THE WAY TO HAPPINESS: PUT DOWN THIS BOOK?

The Happiness Activity Chart is only the beginning of what MacKerron, Mourato, and other researchers working with Mappiness data have told us about happiness. Have you ever wondered:

- » How being a sports fan affects your happiness
- » How substances affect your happiness

» How nature affects your happiness
» How weather affects your happiness

The Mappiness project has given us groundbreaking insights into all of these questions. I am thus going to devote one more chapter, the final chapter of this book, to more lessons from Mappiness and similar modern happiness studies.

But before we get to all of that, I must give you a warning.

You may have noticed the relatively low rank for "reading" on the Happiness Activity Chart. Indeed, this was another activity that Greenberg's and my study suggested people overrate the happiness of.

This is a book that is supposed to give you data-driven life advice, even if it might conflict with the interests of the writer of this book. There is only one more chapter to this book, and I would love it if you read it. But I can't lie. The data says that, if you close this book and call a friend, you are likely to get a happiness boost and probably a bigger happiness boost than you would expect. The data, in other words, suggests you will be happier than you expect if you stop reading my book.

And when you call that friend, you may not want to recommend that they read this cool book, *Don't Trust Your Gut*, on how you can use data to make better life decisions. Instead, you may want to recommend that they go gardening.

That said, if you want to forfeit the 6.38 points of happiness you could get from calling a friend and settle for your 1.47 points of happiness from reading on, you can learn some more about what tends to make human beings happy. And if you want to tell a friend to read *Don't Trust Your Gut* instead of going gardening—knowingly costing them 6.36 points of happiness—I, at least, won't consider you a bad friend.

UP NEXT

Modern happiness datasets can be used to do a lot more than just tell us the average happiness of different activities. We will explore more detailed studies on what tends to make people happy—and miserable.

CHAPTER 9

THE MISERY-INDUCING TRAPS OF MODERN LIFE

"Everything is amazing, and nobody is happy."

This popular saying appears to have first been used as a song title by the Anchors in 2012. It was later popularized in a bit by a now-disgraced comedian on the show *Conan,* featured as the title of an NPR article by Adam Frank, and made available in all sizes as a popular T-shirt quote.

What does the data say?

Well, the quote clearly isn't literally true. *Everything* isn't amazing. Life still brings with it plenty of annoyances for everybody and real struggles for many people.

The amazing blogger and psychiatrist Scott Alexander wrote a depressing and thought-provoking post on a part of modern life that remains un-amazing: the surprisingly high prevalence of people with severe problems in America—real trauma, along with major financial and legal troubles.

Alexander was struck by how many of his patients are in awful situations—people like a seventy-year-old patient with no friends, dwindling savings, and failing health. Alexander wondered how common such objectively bad situations are.

Of course, Alexander realizes that a psychiatrist such as himself gets a biased view of humanity. People with big problems seek out psychiatrists. People without big problems don't seek out psychiatrists. The average person in a psychiatrist's office is likely to be more messed up than the average person.

But Alexander notes that many of us non-psychiatrists also get a distorted view of people, a bias that moves in the opposite direction. People with big problems frequently don't socialize much; some don't even leave their house. The average person in your social circle is likely to be less messed up than the average person.

So how many people in America have severe problems? Alexander looked through data. He found that, at any given time, something like 20 percent of Americans are in chronic pain; 10 percent are dealing with trauma of sexual abuse; 7 percent have depression; 7 percent are alcoholic; 2 percent are cognitively disabled; and 1 percent are in prison. Alexander did some analyses that suggest something like half of Americans, at a given time, may have a severe problem. Alexander concludes, "The world is almost certainly a much worse place than any of us want to admit."

I did my own analysis based on my own expertise—search data—and it validated Alexander's point that many people are dealing with heavy stuff. I analyzed a dataset released by AOL that showed anonymous individuals' search strings over time. I examined the search strings of people who searched for suicide. The strings were heartbreaking and an important re-

Search	Date, Time
looking for a room to rent	March 2, 14:27:12
i need a job	March 2, 15:02:10
seniors	March 2, 23:26:45
www.plentyoffish.com	March 3, 11:18:33
i need a job	March 3, 17:32:00
marriage	March 3, 17:32:31
depression	March 3, 17:33:39
at 60 is life worth living	March 4, 16:43:55
i am being evicted	March 4, 16:57:49
cheap apartment wanted	March 4, 17:00:44
where is the cheapest place to live	March 4, 17:06:32
www.nyclottery.gov	March 5, 16:11:19
poor seniors	March 6, 15:49:04
www.plentyoffish.com	March 6, 20:50:39
www.plentyoffish.com	March 6, 20:51:02
www.plentyoffish.com	March 7, 10:10:53
www.plentyoffish.com	March 7, 10:11:03
christianmingle	March 7, 10:14:00
suicide	March 7, 10:20:36
drugs	March 7, 10:26:27
how to commit suicide	March 7, 10:34:34

minder of the struggles that many people are under, struggles that are frequently hidden from us.*

Consider the searches above, from a senior citizen who is running out of money, being evicted, suffering from loneliness, and struggling to find a job.

---

* *These search strings brought back painful memories for me, as well. I lost more than a decade of my life to severe—and, at times, suicidal—depression.*

The data also reveals that some people have a single problem that they might not tell anybody but that might make their lives so unbearable that they consider ending them. Consider the series of searches below, which is also heartbreaking, of a person in unrelenting chronic pain.

I don't have too much self-help to offer based on these search strings except perhaps to echo this important life advice: "You never know what someone is going through. Be kind." If another person is doing something that is pissing you off, perhaps look at one of these search strings and imagine that the person, when they returned home, made searches such as those. You might find yourself feeling compassion instead of anger.

We must keep in mind that many people live lives that aren't amazing and are, in fact, traumatic. Many people are dealing with heavy stuff.

Search	Date, Time
i can't stand the neck and back pain	April 21, 23:40:05
how can one live with back pain all their lives	April 21, 23:51:45
so depressed because i feel like i have fibromyalgia	May 8, 0:58:43
please help me—i have fibromyalgia	May 11, 1:04:03
is there any hope for fibromyalgia	May 15, 0:57:50
suicide and fibromyalgia	May 15, 0:47:48
i hurt so much from arthritis and tmj	May 18, 13:30:21
pain at bottom of neck and top of back	May 19, 22:24:21
help with pain from arthritis and fibromyalgia	May 19, 0:26:51
back neck hurts	May 20, 11:17:58
miserable from neck pain back pain and tmj	May 20, 0:18:02
suicide	May 23, 12:13:05

Furthermore, of course it isn't literally true that "*nobody* is happy." In fact, according to the General Social Survey (GSS), 31 percent of Americans rate themselves as "very happy" these days.

But, if "everything is amazing right now, and nobody is happy" isn't literally true, it is directionally true. Even if life isn't amazing for everyone, life has consistently gotten *way* more amazing by many metrics. Yet, in general, the increased number of people in increasingly amazing objective situations hasn't translated to increased happiness.

First, the amazingness data.

In the past fifty years, the gross domestic product (GDP) per capita in the United States, even when you control for inflation, has more than doubled. Pretty amazing!

What's more, GDP captures only the value of goods and services that people buy. Our digital economy gives us many

Everything is amazing:
Real GDP per capita in United States, 1972-2018

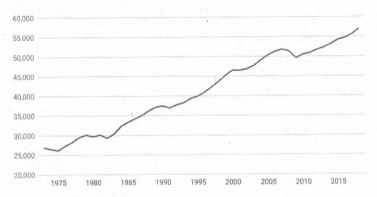

Source: U.S. Bureau of Economic Analysis · Created with Datawrapper

things for free, things that won't show up in GDP. A recent analysis tried to figure out how much some of these services are worth to people by asking them how much they'd have to be paid to give them up. They estimated that search engines are worth $17,530 every year to the average American; email is worth $8,414; digital maps $3,648; and social media $322. We pay $0 for these services. Pretty amazing!

Next, the happiness data. Over this same period, re-ported happiness, at least in the United States, hasn't gone up. In 1972, the first year the GSS collected data, when GDP per capita was less than half as high and nobody had Google or Google Maps or Gmail, 30 percent of Americans said they were "very happy," roughly identical to the number today.

So, it is true that, despite things being increasingly amaz-ing, people are not getting noticeably happier. Why?

People aren't happier:
% of Americans who reported they were "very happy," 1972-2018

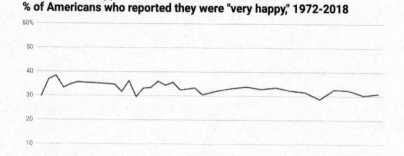

Source: General Social Survey • Created with Datawrapper

One reason that we aren't getting happier even though we are getting richer is that money has only a small effect on happiness. Matthew Killingsworth is the founder of track yourhappiness.org, a project that, like Mappiness, pings iPhone users and asks them to report their happiness. He conducted the largest study ever on the relationship between individual income and happiness—a study of roughly 1.7 million data points. He found that income does increase happiness, but the effects are small, particularly for high incomes. Doubling your income can be expected to increase your happiness by about one-tenth of a standard deviation, which really isn't much.

Another challenge in the human quest for happiness is that our minds kind of suck. A friend of mine once said that, if the human mind were an operating system, the owner would send it back, complaining that it was buggy. One crucial bug in our mind that limits our happiness: our inability to focus on the present moment.

This was shown in another study by Killingsworth, this time in collaboration with Daniel Gilbert. The researchers, in addition to asking their trackyourhappiness.org users what they were doing and how happy they were, asked them the following question: "Are you thinking about something other than what you're currently doing?" (They also asked whether the other thought was pleasant, neutral, or unpleasant.)

They found that 46.9 percent of the time, a person is thinking about something other than what they are currently doing. The researchers also found that, when a person is thinking about something else, they are significantly less happy.* Shockingly, even when a person's mind is wandering in a pleasant direction, they report being slightly less happy than if they are focused on the task at hand. And, if the wandering thought is neutral or unpleasant, people are miserable.

As the authors summed up their research, "A human mind is a wandering mind, and a wandering mind is an unhappy mind."

The dangers of a wandering mind may be part of the reason meditation can prove so helpful for happiness. Scientists have consistently found that meditation does boost happiness.

The buggy operating system that is the human mind is a big part of the reason happiness can be such a challenge, even in an increasingly amazing world. But there are other reasons that people struggle to be happy in modern times, many of which have been revealed by the Mappiness project. Quite simply, people spend a lot of time in situations that are unlikely to make them happy.

Since 2003, the Bureau of Labor Statistics has conducted the American Time Use Survey, which asks a sample of Americans to report how they spend every moment of a day.

I compared the data from the American Time Use Survey to the Happiness Activity Chart created by Mappiness, which I discussed in the previous chapter. I divided activities into three categories: those that give the most happiness, everything from making love to meditating and religious activities; those that give medium happiness, everything from playing with children

---

* Interestingly, sex was the only activity in which fewer than 30 percent of participants were thinking about something else, which may be part of the reason it is the happiest activity.

to listening to a podcast; and those that give the least happiness, everything from washing and grooming to being sick in bed.

I found that the average American spends only about 2 hours per day doing one of the happiest activities. In contrast, they spend some 16.7 hours per day doing one of the least happy activities. Some of this, admittedly, is due to sleep counting as one of the least happy activities. And the average American sleeps roughly 8.8 hours per day. But even taking out sleep, Americans spend about half of their waking hours doing things that rank among the least happy activities—predominantly working, housework, commuting, and grooming.

Further, Americans do not seem to be translating their increased wealth into more time doing things that tend to make people happy. Between 2003 and 2019, a period in which real per capita income rose more than 20 percent and Silicon Valley delivered all these amazing free products, I found that Americans, if anything, spent less time doing the happiest activities. This was predominantly due to a decline in time spent "talking/chatting/socializing" from 0.93 hours to 0.77 hours, as well as small declines in time spent on "meditating/religious activity" and "gardening."

The inability to translate wealth into time spent doing things that make us happy points to a larger problem with mod-

AVERAGE HOURS SPENT PER DAY

	Least Happy Activities (e.g., Working/ Housework)	Medium Happy Activities (e.g., Eating/Childcare)	Happiest Activities (e.g., Socializing/ Theater)
2003	16.71	5.22	2.07
2019	16.72	5.42	1.86

ern life, revealed by more studies using Mappiness and similar data sources. Modern life, quite simply, puts traps in front of us that can get in the way of our happiness. And, if you can avoid them, you are far more likely to be happy.

THE WORK TRAP

Work sucks.

That is perhaps the most striking fact in the Happiness Activity Chart. Work ranks as the second most miserable activity, besting only being sick in bed.

This isn't always what our acquaintances tell us. When you ask other people about work—at cocktail parties, at networking events, on social media—some will make statements such as "I live for my work," "I love my job," or at least "I like my job."

But when Mappiness asks people how much they are enjoying themselves, at the moment they are at work, with no other human beings able to see their answer, people report a more grim view of work. They report that work, on average, makes them more miserable than doing chores, taking care of old people, or waiting in line. This suggests many people may be lying—to themselves or to you—when they report that they like or love their job.

The misery that so many people feel at work is, in fact, quite sad. Think about it. The majority of adults spend a large fraction of their waking hours at work. If most of these people, most of the time that they are at work, feel miserable, this implies that most adults spend a large portion of their waking hours deeply unhappy.

There is no obvious solution to the work problem. Some findings from the Happiness Activity Chart have obvious solutions.

PROBLEM: Happiness Activity Chart reveals that commuting makes people unhappy.
SOLUTION: Live closer to your job.

PROBLEM: Happiness Activity Chart reveals that smoking makes people unhappy.
SOLUTION: Quit smoking.

But most adults can't quit working. Most adults need to work many hours to feed and clothe themselves and their family members.

Does this mean that adult life is destined to stink, leaving only a few hours per week for activities that people enjoy?

Not necessarily.

Yes, work, the predominant activity of adult life, is frequently miserable. Mappiness has shown that. But there are ways to make work less miserable—and even enjoyable. There are some people who find work less painful than others do.

So, what explains why some people find work tolerable and occasionally fun? And how can you be one of those people?

George MacKerron, cofounder of the Mappiness project, studied what else people did while they worked—and what seems to make people enjoy work at least a bit more. On this project, MacKerron worked . . . with his friend Alex Bryson. That was a wise decision for boosting happiness. More on that in a moment!

The first thing that makes work a bit less miserable, MacKerron and Bryson discovered: listening to music. About 5.6 percent of the time people are working, they are also listening to music. And people who are listening to music as they work report a boost of happiness of 3.94 points, bringing their overall happiness score while they work to –2.66. If you find your job mind-numbing, try to get some feeling back in your mind by listening to some tunes.

The second thing that makes work a bit less miserable: working from home. MacKerron and Bryson found that people working from home were, on average, 3.59 points happier. That said, someone listening to music, while working from home, is still likely to be fairly miserable.

The Mappiness data revealed only one way to truly make work tolerable—or even enjoyable. This is the third and most important data-driven way to lower the sting of work. In the Mappiness dataset, individuals who were at work with people they considered their friends were far happier than others at work. Being with friends can give such a large boost in happiness that it may make work a pleasurable experience. People can turn the daily, painful grind into the daily, pleasurable grind with a little help from their friends.*

THE WORKINGMAN'S BLUES (OR LIGHT BLUES)

Working (Base Case)	–5.43
Working at Home	+3.59
Working While Listening to Music	+3.94
Working with Your Friends	+6.25

Source: Author's calculations from Bryson and MacKerron (2017)

---

* *This is a somewhat technical point. In MacKerron and Bryson's paper, the boost from working with friends consists of the average effect on happiness of being with friends plus the interaction effect of being with friends while working. The bonus of working with friends comes exclusively from the average effect on happiness of being with friends. This means that friends always give a bonus to happiness, including while you work. I explore this further in the next section.*

My estimate, based on MacKerron and Bryson's numbers, imply that the average person working with their friends is about as happy as they would be if they were relaxing alone. Recall from the previous chapter that relaxing doesn't produce as much happiness as people would have guessed. But it is still far better than the misery of the average worker.

The average person working from home, listening to music, with their friends (perhaps on Zoom or perhaps because the friend had come over for a work session), would be about as happy as the average person playing sports—which, you may recall, is one of the single happiest activities.

ALBERT CAMUS FAMOUSLY WROTE ABOUT THE GREEK LEGEND OF Sisyphus, who is condemned by the gods to repeatedly roll a boulder up a hill. When Sisyphus reaches the top of the hill, the boulder rolls down, forcing him to do it again. Camus thought this story was a metaphor for the modern workingman, who must spend his working life repeatedly performing pointless tasks. Sisyphus had to roll a rock. Dunder-Mifflin employees have to write memos.

This seems like a dark vision of modern life. But Camus ends his essay with a twist. He writes, "All is well." The final words of his essay: "One must imagine Sisyphus happy." With one shocking paragraph, a pessimistic and brutal essay is transformed into an optimistic tale.

Modern data tells us that Camus, like so many other renowned philosophers who pontificated without proper tools of measurement, while he may have been clever, was dead wrong. All is not well for modern workers. If we imagine that they are happy, we are mistaken. In fact, if we want our imaginations to correspond to reality, we must imagine that the Sisyphuses among us are experiencing 5.43 points of misery.

If a modern philosopher wants to write a metaphorical story with some basis in reality, he might try *The Myth of Sisyphus and Sisyphas*, featuring a pair of best friends Sisyphus and Sisyphas who are condemned to repeatedly roll a boulder up a hill—together.

Sometimes, they push the rock as a team. Sometimes, one person pushes the rock as the other rests. Sometimes, some idiot with more power tries to teach them to push the rock a better way, while they roll their eyes and mock the idiot. Sometimes, they both take a break from rock pushing to discuss their dating lives, favorite TV shows, or fantasy football rosters. All is well. Sisyphus and Sisyphas are, according to the best tools human beings have ever invented to measure our moment-by-moment utility, indeed happy.

Or, to summarize the data on the world of work: Pay a lot of attention to whom you work with. If you work with friends, work is far more likely to be pleasurable.*

---

* *Another data-driven life hack related to work is to quit a job that sucks. In an incredibly clever study, Steven Levitt, a coauthor of* Freakonomics, *asked people who were facing a major decision—such as whether to quit a job—to do it by way of a coin flip. Incredibly, many people were willing to follow the advice of the coin. Levitt found that, months later, people who were induced by the coin flip to quit their job reported being significantly happier than those who were induced by the coin flip to stay at it.*

THE "NOT SPENDING ENOUGH TIME WITH FRIENDS AND ROMANTIC PARTNER" TRAP

Friends are the key to happiness in many dimensions of life, not just one's work life. In fact, the happiness boost from friends isn't particularly unusual among those working. In another paper, MacKerron, again smartly working with a friend, Susana Mourato, examined how being around other people affects happiness.

In this paper, the researchers compared the same people, doing the same activity, at the same time of day—but whether they were by themselves or with other people. If they were with other people, the researchers compared being with different types of other people—romantic partners, friends, family members, etc.

The results?

The people who make us happiest are those we choose: romantic partners and friends. Relative to being by themselves, people, on average, gain more than four points of happiness being with a romantic partner or friends.

Other people, however, tend not to make us happy. On average, people gain only a little bit of happiness—or would be happier alone—when they are with people other than a romantic partner or friends.

HAPPINESS PEOPLE CHART

Person	Gain in happiness from being with them, relative to being alone
Romantic partner	4.51
Friends	4.38
Other family members	0.75
Clients, customers	0.43
Children	0.27
Colleagues, classmates	−0.29
Other people participant knows	−0.83

Source: MacKerron and Mourato (2013)

People need other people to get by, we are often told. People are naturally social creatures, we are told. And clearly, we *can* be a lot happier with other people. But the research of MacKerron and Mourato shows that the happiness boost depends mightily on who these other people are. We tend to be less happy with many people we don't know that well—or didn't choose to enter our close social circle.

We get a big happiness boost interacting with our romantic partner or our close friends. But random old classmates? Work colleagues? People we sort of, kind of know? Interacting with these people, the data suggests, tends to not make us happy. In fact, the data tells us that we are frequently happier by ourselves than when interacting with many of our weak ties. Or, as George Washington reportedly said, "'Tis better to be alone than in bad company." Or, as George Washington might have said, had he lived long enough to learn about modern happiness research, "On a 0–100 scale, 'tis 0.83 points better to be alone than in bad company."

THE SOCIAL MEDIA TRAP

Does social media make us miserable?

Yes.

The Happiness People Chart suggests that social media might make us unhappy. Our time on social media is not merely spent interacting with our romantic partner or close friends—the people who tend to make us happy. We also interact with weak ties—the people who tend not to make us happy.

The Happiness Activity Chart also suggests that social media makes us unhappy. Social media scores as the single lowest happiness-producing leisure activity.

But we have more evidence than that.

Researchers from New York University and Stanford University recently ran a randomized controlled experiment on the effects of using Facebook. These researchers divided people into two groups—a treatment and a control group. One group of users, the treatment group, was paid $102 each to stop using Facebook for four weeks.* The other group of users, the control group, was allowed to proceed with their lives as normal.

More than 90 percent of people in the treatment group indeed stopped using Facebook. So, what happened when people did that?

Compared to the individuals in the control group (who continued using Facebook as before) people in the treatment group (who signed off Facebook) spent sixty minutes less on

---

* *They chose this number because it was the average amount people reported that they would need to be paid to stop using Facebook.*

social media, with much of the saved time spent with friends and family. And these people reported that they were happier. The gain in happiness from not using Facebook was about 25–40 percent as large as the gain in well-being from entering individual therapy.

In addition, most people, after the intervention, realized that they were happier. About 80 percent said that the deactivation was good for them. Following the experiment, they succeeded in using Facebook less in the following month.

Of course, most of us are not being offered $102 to quit using Facebook for four weeks. But we can take advantage of the lessons from those who were—and cut down on our use of Facebook and similar social media platforms. It is, the data tells us, making us miserable.

THE SPORTS TRAP

I really, really, really, really, really *love* sports! You may have guessed that from the fact that I talked about my Mets obsession in the intro or that I consider this book "Moneyball for your life" or that, in a book that is meant to explore the nine most fundamental questions of life, one chapter explores how to be a world-class athlete.

But, yeah, I'm a huge sports fan. Always have been. I think I always will be.

So, it is with absolutely zero pleasure that I, certified sports nut, ask the following question: does watching sports make people miserable?

An extremely important study by MacKerron, with Peter Dolton, of the University of Sussex, has certainly made me rethink the (enormous) role that sports plays in my life. Mac-Kerron and Dolton wanted to see how sports fans' happiness was affected in the hours after their favorite teams won or lost a match.

For a group of fans of various soccer teams, MacKerron and Dolton studied the average fan's minute-by-minute happiness before, during, and after their favorite team's matches. The results?

Let's start with what happens before the match starts. In the minutes before a match begins, the average sports fan gets a slight boost in happiness (about 1 point on the 100-point scale). The average fan perhaps anticipates a win and gets pleasure imagining this win.

Next, what happens after the match? Not surprisingly, it depends on what happened during the match.

If a fan's team wins, the fan gets an additional boost of about 3.9 points of happiness. Not bad! So far, so good for being a sports fan. If your sports team wins, being a sports fan is fun.

But what happens when your sports team loses? If a fan's sports team loses, they can expect to lose 7.8 points of happiness. (A draw gives the average fan 3.2 points of pain.) In other words, losses hurt the average sports fan far more than wins please them.

Sports fans seem to have made a terrible bargain. Since, on average, teams can expect to win as many games as they lose, a sports fan can expect to suffer more than he rejoices.

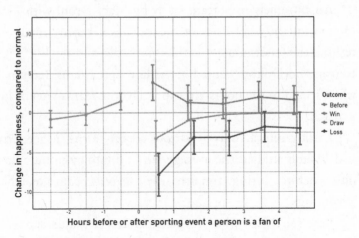

The Sports Fan Trap: The Contentment of Victory, the Agony of Defeat

MacKerron kindly sent figures and charts from this working paper.

These effects are large. Suppose that somebody supports four teams. Maybe he is a fan of the Knicks, Mets, Jets, and Rangers. Over the course of a year, and extrapolating from the Mappiness results, he might expect to lose 684 points of happiness. Being a passionate fan of four sports teams, in other words, is roughly the equivalent for one's mood of being sick in bed an extra 2.2 days every year.

So what is the sports fan to do? Is there any way to escape the Sports Trap?

One obvious option is to support better teams. The math seems to say the following: if you get 3.9 points of pleasure from a win and 7.8 points of pain for a loss, then as long as your team wins more than 66.7 percent of its games, you will get more pleasure than pain from supporting that team.

My father attempted to do this. After years of frustration caused by his support of the New York Mets, who just about always stink, he decided to switch his allegiance to the New York Yankees, who frequently compete for championships. As my dad put it to me on one crisp fall evening, "Son, life's too short to root for a shitty team."

Andrew Yang, the entrepreneur and politician, made a similar calculation in basketball fandom. He switched his life-long allegiance from the New York Knicks to the Brooklyn Nets. "Better than the Knicks, man," he told *Forbes*. "The ownership is just too . . . too shitty."

Have my dad and Yang outsmarted the system? Have they escaped the Sports Trap that MacKerron and Dolton discovered?

Nope!

MacKerron and Dolton sliced the data further. They found that the sports fan's brain adjusts to how good their team is, limiting how much pleasure they can get from the wins of a great team. In particular, the researchers found that a sports fan, when his team is expected to win the game, will get only 3.1 points of pleasure from a win and will lose 10 points of happiness from a loss. In other words, the better the team you support is, the more the team has to win to give you any pleasure.

	Team Was Expected to Win	Team Was Expected to Lose
Average Happiness Change After a Win	+3.1	+7.0
Average Happiness Change After a Loss	−10.0	−6.3

A similarity to many addictions might be noted here. For many substances—think, for example, of cocaine—the more you have consumed it, the less additional consumption can give you pleasure—and the more any breaks in consumption will give you pain.

As with cocaine, so with the Yankees. A team that has won a lot has to win that much more to give you pleasure—and any break from the expected winning will cause you extreme pain. The data implies that it is simply impossible to find a sports team that wins enough to escape the pleasure-pain trap of sports fandom.

Does this mean we sports fans should cut out sports completely from our lives? Is the Mappiness data the equivalent of those early studies that showed that smoking causes cancer? Should sports games have a Surgeon General's warning on them, telling us that we can expect them to make us miserable?

Not exactly.

Return to the Happiness Activity Chart from a few pages earlier. (And if you are too lazy to go back pages and check, I will just tell you what it says.) You will note that watching sports, on average, can be a reasonably pleasurable experience, ranking in between hobbies and playing with pets.

The danger of watching sports, from a happiness perspective, does not arise with every single game. Instead the danger comes when we are fans of the team—in other words, when we care too much about the outcome. The data suggests that we have a greater chance of enjoying watching sports the less we care about the outcome.

We all need to be more Buddhist in our sports consumption. When we watch sports and don't care about the outcome so much, we can appreciate the artistry of world-class athletes. When we watch sports and do care about the outcome, we run into a trap of losses hurting us more than wins nourish us.

Or, to summarize the data from the world of sports: watch more sporting events featuring teams you're not fans of.

THE BOOZE TRAP

Neal Brennan, a comedian who has struggled with depression his entire life, was once given advice from his friend the comedian Dave Chappelle. Chappelle advised Brennan to fight his blues the following way: "Just drink." "But I don't, I don't like drinking," Brennan said. Chappelle was noting that many adults—though not his friend Brennan—use substances, such as alcohol, to cure the blues of adult life.

Is this wise?

The sober advice, of course, is the opposite of Chappelle's. Many people are told to avoid substances, to figure out how to be happy without artificial mood-enhancers. This is undoubtedly the right advice for the significant fraction of the population that gets addicted to alcohol. Alcohol is a dangerous substance that can ruin people's lives.

But what about non-addicts? Should they drink? When should they drink?

Ultimately, there are many relevant empirical questions here. How happy does drinking actually make people? Do people pay a price after they drink, in a lowered mood in the hours or days following? Does the mood when people drink depend on what else they are doing?

The honest answer to these questions ten years ago would

have been, respectively, "no clue," "no clue," and "no clue." But Mappiness, as with so many other issues related to happiness, has transformed our understanding of alcohol's role in happiness. Mappiness users were able to report whether they were "drinking alcohol" and how happy they were. MacKerron and a coauthor, Ben Baumberg Geiger, studied the data on the happiness of booze.

First, the not so shocking. The same person, doing the same activity, with the same people, will be about four points happier if he or she is also consuming alcohol. Alcohol really does make people feel better.

Now, what happens afterward? Does a person lose the four points they gained? Does Sunday morning take away what Saturday night gives? The researchers tracked people after they drank. They found, on average, no difference in the mood the morning after an evening with alcohol. They did, however, find that people who drank were slightly more tired the following morning than people who did not drink.

The researchers could also break down the booze boost by different activities. They looked at two questions: What activities do people tend to supplement with alcohol? How much does alcohol boost happiness for different activities?

Let's start with the first question. People are most likely to drink, not surprisingly, when they are socializing. One striking pattern the scientists found is that people are most likely to drink when they are doing an activity that is already fun, with or without alcohol. People drink, in other words, to try to turn a great night into an epic night.

Now, to the second question: how does booze affect our

WHAT PEOPLE ARE DOING WHEN THEY DRINK

Talking/Chatting/Socializing	49.2%
Watching TV/Film	31.2%
Eating/Snacking	27.9%
Listening to Music	10.4%
Sleeping/Resting/Relaxing	7.4%

Source: Geiger and MacKerron (2016)

mood when engaged in different activities? It turns out, we tend to get the biggest mood boost from alcohol when we are doing an otherwise unpleasant activity.

A Bruce Springsteen concert is really fun, with or without alcohol. Having sex is really fun, with or without alcohol. Talking to your friends is really fun, with or without alcohol. But traveling/commuting tends to suck without alcohol, while it is okay with alcohol. Ditto for waiting/queueing and washing/dressing/grooming.

These results suggest that many of us use alcohol incorrectly, as we are most likely to drink during some of the activities that offer the least mood boost from alcohol. The results

Biggest Booze Boost (People Are MUCH Happier Doing These Tipsy Than Sober)	Smallest Booze Boost (People Are Roughly as Happy Doing These Tipsy as Sober)
• Traveling/Commuting	• Intimacy/Making Love
• Waiting/Queueing	• Theater/Dance/Concert
• Sleeping/Resting/Relaxing	• Talking/Chatting/Socializing
• Smoking	• Watching TV/Film
• Washing/Dressing/Grooming	• Reading

Source: Geiger and MacKerron (2016)

also suggest some counterintuitive drinking strategies that could actually improve your well-being.

Say, for example, you are getting ready for a night out with your friends. Most people would be sober as they are getting ready and would drink once they were out. The data suggests you might actually be happier if you had a drink in the shower, were a little tipsy as you got ready, and then sobered up when you were out. You would have fun, tipsy, as you got ready and fun, sober, when you were out.

Say, for another example, you are going to a concert and then taking an Uber back. Most people would have a couple of drinks at the concert and be sobered up for the boring Uber trip back. The data suggests you might have a happier night if you stayed sober for the concert and had a couple of drinks just before you got the Uber back. You would have fun, sober, at the concert and fun, tipsy, on the commute.

Again, I need to emphasize the dangers of alcohol. Alcohol ruins some people's lives, and, of course, you need to be cautious if you are going to drink with your colleagues, drink in the shower, or drink on planes.

So, I will offer the advice with a crucial caveat. If you do not have an addictive personality, there is evidence that alcohol can be an important mood-enhancer. And you might consider using it less when you are doing something already fun, such as socializing or having sex, and more to help make painful and boring activities less painful and boring. But you must, of course, be extremely careful with this advice. There is a thin line between counterintuitive, data-driven strategies to improve your mood and a sure path to alcoholism.

THE NOT-SPENDING-ENOUGH-TIME-IN-NATURE TRAP

"Happiness Is Greater in Natural Environments." This is the title of a paper by MacKerron and Mourato based on their analysis of Mappiness data.

The paper argues (as you may have guessed from the title) that being in nature is a crucial component of happiness. Want to be happier? The scientists advise us to try to spend more time in fields, mountains, and lakes—and less time in subways, conference rooms, and couches (unless, of course, you are having sex on the couch, which, as mentioned, is the most pleasure-producing activity in which human beings can engage).

How do the scientists make their claim that being in nature makes people happy?

An apparent relationship between being in nature and well-being is striking just looking at the Happiness Activity Chart. Five of the ten activities (50 percent) that people are most happy doing—playing sports, gardening, birdwatching, hunting/fishing, and walking/hiking—are frequently done in natural environments. Ten of the ten activities (100 percent) that people are least happy doing are virtually never done in nature.

Of course, a *correlation* between being in nature and being happy does not, by itself, prove that being in nature *causes* people to be happy. Perhaps the reason activities done in nature are associated with more happiness is that the activities happen to be more fun, not because of where people are doing them.

Sure, the most miserable activity—being sick in bed—is never done in nature. But most people would be miserable even if they were sick in a natural environment. If you were lying at the edge of the Grand Canyon at sundown, but your throat was burning and your stomach aching and your head throbbing, you likely would still be miserable.

The evidence for the causal impact of nature on happiness, however, gets substantially stronger. MacKerron and Mourato didn't merely compare all people who were doing something in nature (such as the people hiking Yosemite on a cloudless Saturday in May) versus all people who were doing something in an artificial environment (such as the people who were sick in bed on a damp Tuesday in February), which would not be all that convincing. They compared the same person in otherwise identical situations—in any way they could think to measure—but in different environments.

Here's how it works. Say John takes a run on Fridays at 5 P.M. with his boyfriend, as long as it is warm and sunny. Some warm, sunny days, he just runs the streets of London. Sometimes he goes to a park. Other times he runs by a lake. They could compare his happiness in each of these occasions. Say Sarah is in work meetings every Monday at 2 P.M. But one time, it is outside in a grassy field, whereas other times it is in a standard room. They could compare her happiness in these different situations. And because Mappiness's dataset is enormous, they can do this for many people.

How do they measure whether a person is in nature? Recall that Mappiness only asks people to report what they are doing and with whom, not where they are doing it. This is the magic of working with iPhones. The iPhones give a GPS report of the person's latitude and longitude. And MacKerron and Mourato could then match that up with another dataset that reports, for every corner of the country, the type of land cover. So, without further ado, below are the results of how happy the same person, doing the same activity, at the same time, is in different types of environments.

And there you have it. After reading this chart, I felt I had finally found, with a huge assist from MacKerron and Mourato, the secret to my problems, the answer to happiness.

If I want contentment, all I have to do is: . . . spend more of my time in "marine and coastal margins." There was only one remaining problem: I hadn't the slightest clue what "marine and coastal margins" meant. Nor, for that matter, did I know what "moors" and "heathland" were; MacKerron and Mourato found they also give a hearty boost of pleasure.

Thankfully, I could Google these terms. It turns out that "marine and coastal margins" means the land near the sea and ocean.

While MacKerron and Mourato didn't study the effects of looking at pictures of "marine and coastal margins," I thought I'd offer a picture—on the next page—just in case that improves people's mood as well.

HAPPINESS GEOGRAPHY CHART

Land Cover	Happiness Gain (compared to being in an urban environment)
Marine and coastal margins	6.02
Mountains, moors, and heathland	2.71
Woodland	2.12
Semi-natural grasslands	2.04
Enclosed farmland	2.03
Freshwater, wetlands, and floodplains	1.8
Suburban/rural developed	0.88
Inland bare ground	0.37

Source: MacKerron and Mourato (2013)

Heathland and moors, according to Wikipedia, are "free-draining infertile, acidic soils . . . characterized by open, low-growing woody vegetation," which I still don't really understand. But a picture of them is on the following page.

The World's Happy Place. Photo was contributed to Shutterstock by ZoranKrstic.

How should we think about the size of the effects of "land cover" on one's happiness? Let's compare them to the Happiness Activity Chart. If you are sitting in a meeting, you can expect –1.5 points in happiness. But if you are doing it in "marine and coastal margins"—for example, in land near water—you can expect about 4.5 points in happiness, which would be roughly equivalent to watching a sporting event. In other words, moving a meeting from a sterile urban conference room to land near water takes a boring activity into an okay activity. Pretty big effect!

One of the striking aspects of nature is that it tends to be beautiful. There is also evidence from Mappiness data that

Being in a Place Like This Gives You 2.71 Happiness Points.
Photo was contributed to Shutterstock by Israel Hervas Bengochea.

being around beauty itself, even if it is not in nature, can improve our mood.

Chanuki Illushka Seresinhe and coauthors explored this further. They utilized a new site, ScenicOrNot, which has asked a team of volunteers to rate the beauty of every corner of Great Britain. The photo on the next page, for example, shows a place that ScenicOrNot volunteers rated as beautiful.

The researchers could use the GPS data on Mappiness users to see what corner of Great Britain they were in—and how beautiful that place was. The authors could then take into account everything MacKerron and Mourato took into account in their study—the activity, the time, the people a person is with, the weather—plus the land cover. They could now compare the same person, doing the same activity, with the same people, at the same time, in the same climate, in the same

This place was rated as beautiful. All else equal, people report being significantly happy in locations such as this. Source: Photo by Bob Jones. Image is licensed for reuse under the Creative Commons Attribution-Share Alike 2.0 Generic License, which can be found at http://creativecommons.org/licenses/by-sa/2.0/.

type of place (for example, "marine and coastal margins"), but surrounded by different levels of beauty.

All else being equal, the researchers found that being in the most scenic places—compared to the least scenic places—adds 2.8 points of happiness. The lesson from the data is clear. A way to be happier is to spend more time in nature and around beauty.

These results were confirmed by a different team, led by Sjerp de Vries. Motivated by the Mappiness project, they created their own app, HappyHier, which pinged Dutch people and asked them how happy they were. They also found that people were happiest in nature and near coasts or water. Interestingly, these researchers found that the happiness effect of being near the coasts also held when a person was indoors—perhaps because the view cheered them up.

ONE FINAL MAJOR QUESTION ABOUT HOW OUR SURROUNDINGS influence us: how does weather affect our happiness? Mac-Kerron and Mourato analyzed this, as well. Once again, the researchers compared the same person, doing the same activity, at the same time—but looked at how their happiness changed as the weather changed.

The direction of the results were not surprising at all. Sun makes people happier than rain. (Duh!) Warm days make people happier than cold days. (Duh!)

But the magnitude of the effects was somewhat surprising. In particular, by far the biggest weather effect on our happiness is

HAPPINESS WEATHER CHART

Weather	Happiness Effect When Participant Is Outdoors
Snow	1.02
Sun	0.46
Fog	−1.35
Rain	−1.37
0–8 degrees C	−0.51
8–16 degrees C	0.29
16–24 degrees C	0.99
24+ degrees C	5.13

Source: MacKerron and Mourato (2013)

the positive effect of really warm days. When it is 24 degrees Celsius (or 75.2 degrees Fahrenheit) or higher, people report, on average, 5.13 points of additional happiness. Other weather effects are, in comparison, minor. People are not much more miserable if it is frigid as opposed to just cold. And the negative effects of rain are significantly smaller than the positive effects of warmth.

As far as weather is concerned, happiness seems more about maximizing the number of perfect weather days than avoiding awful weather days.

Comparing the Weather Happiness Chart to other charts that MacKerron, Mourato, and the Mappiness team have discovered also leads to an important point: frequently, other factors matter more than the weather in determining one's happiness.

For example:

» Even while outside, people are, on average, happier hanging out with friends on a rainy, 35-degree day than being by themselves on a sunny, 70-degree day.
» People are happier being by a lake on a 35-degree day than being in a city on a 70-degree day.
» People are happier drinking alcohol on a 35-degree day than being sober on a 70-degree day.
» People are happier playing sports on a 35-degree and rainy day than lying down and doing nothing on a 70-degree, sunny day.

Perfect, sunny days can indeed improve our mood. But it's important not to exaggerate weather's importance. Weather, by itself, can't make you happy. You also need to be doing things that make you happy with people who make you happy.

CONCLUSION

Readers, it has come time to sum up this book. And I know I'd better make it a good summary because of the peak-end rule discussed in Chapter 8. Your feelings about this entire book are going to depend heavily on how you feel about these final few paragraphs. And, for those of you who think that reading this book felt like having a colonoscopy, perhaps I can at least make it like Patient B's colonoscopy in Chapter 8: not so bad in the end, allowing you to remember the experience as less painful.

So, what have we learned from data available from dating sites, tax records, Wikipedia entries, Google searches, and other sources of Big Data?

Well, Big Data shows us that we often have notions about the way the world works that are different from the way it actually works.

Sometimes the data reveals truly counterintuitive insights, such as that a typical rich American is the owner of a wholesale beverage distribution company or that couples with wildly different qualities are equally likely to get more happy or more miserable over time.

Sometimes, however, the data reveals counter-counterintuitive insights. These insights make plenty of sense but somehow have not become conventional wisdom. Unrepresentative data from the media and other sources of information in modern life have plainly tricked us.

This is among the major lessons of the research of George MacKerron, Susana Mourato, and others on probably life's most important topic: happiness. Reading the groundbreaking modern studies on happiness, I came to the conclusion that happiness is less complicated than we sometimes think. The things that tend to make people happy—say, hanging out with friends or walking near a lake—aren't exactly mind-blowing.

Yet modern society tries to fool us into doing things that data (or even a little common sense) says are unlikely to make us happy. Many of us devote years working far too hard at jobs we don't like with people we don't like. Many of us spend hours poring over the latest updates on social media. Many of us go months without spending real time in nature.

Mappiness data—and similar projects—tell us that, if we are not happy, we have to ask ourselves if we are doing enough of the (not exactly earth-shattering) things that tend to make human beings happy.

After reading through all the happiness studies, I asked myself if we could boil down all the lessons from modern happiness research into one sentence. We might call such a sentence "the data-driven answer to life."

How might we sum up what Big Data tells us about life's most important question? What do millions of pings only available thanks to smartphones reveal about the answer to the mystery of suffering and existence? What, more generally, is the data-driven answer to life?

The data-driven answer to life is as follows: be with your love, on an 80-degree and sunny day, overlooking a beautiful body of water, having sex.

ACKNOWLEDGMENTS

Frequently, when I read a book, I read the acknowledgments first. Am I the only one who does this? Anyway, I hope any fellow acknowledgment aficionados enjoy what follows.

My biggest thanks are to the scientists discussed in the book. I thank them both for their work and for talking to me about it. In particular, I benefited from discussions with Albert-László Barabási, Paul Eastwick, Sam Fraiberger, Samantha Joel, George MacKerron, Alexander Todorov, Danny Yagan, and Eric Zwick.

Some of my interpretations of the work may be a bit different from those of the scientists who conducted the studies; all of the original studies can be found in the endnotes.

For help with gathering data and stories and collaborating on research projects, I thank Anna Gát, Spencer Greenberg, David Kestenbaum, Lou Corina Lacambra, and Bill Mallon.

For offering feedback on sections, I thank Coren Apicella, Sam Asher, Esther Davidowitz, Amanda Gordon, Nate Hilger, Maxim Massenkoff, Aurélie Ouss, Julia Rubalevskaya, John Sillings, Katia Sobolski, Joel Stein, Mitchell Stephens,

Lauren Stephens-Davidowitz, Noah Stephens-Davidowitz, Logan Ury, and Jean Yang.

Thanks to Sourav Choudhary and Adam Shapiro for offering me consulting opportunities that came with friendship and gentle nudges to finish my book.

For more new friendships and gentle nudges to finish my book, I thank the Hirsch and Seessel families.

For less-gentle but more effective nudges to finish my book, I thank Matt Harper. Matt is a wonderful editor who had the unenviable task of keeping me focused—and handled it splendidly.

For impeding me from finishing my book with endless memes and ferocious political debate, I thank #YouAreFake News.

Melvis Acosta is a rock star fact-checker with an attention to detail that I did not realize was humanly possible. Any remaining errors are likely due to my skipping over one of the items in the pages of notes Melvis sent on every chapter.

Eric Lupfer remains an off-the-charts thoughtful and creative agent.

The research in chapter 2 says that parents can influence how their kids think about them. And I think my parents are the best in the world. So causal outcome achieved, Mom and Dad! I also suspect we are an outlier family in how much you have helped my career.

The research in chapter 9 says that people aren't, on average, much happier when spending time with their family. But if Mappiness or another experience sampling service measured my happiness, I am confident that there would be a big boost when I am with Noah, Lauren, Mark, Jonah, Sasha, and the rest of the Stephens-Davidowitz-Osmond-Fryman-Wild-Sklaire clan.

And if Mappiness followed me for the past decade, they'd notice a dramatic change in my mood after working with the world's greatest therapist, Rick Rubens. Thank you, Rick, for helping me work through my depression.

Julia, thanks for everything. You know I struggle with expressing emotional warmth; but you also know how much I love you.

APPENDIX

The chart that follows compares the predicted happiness of activities, according to the survey conducted by Spencer Greenberg and me, to the actual rank of activities, as found by Bryson and MacKerron. Activities that have a positive Difference, such as "Exhibition/Museum/Library," tend to give people more happiness than people expect. Activities that have a negative Difference, such as "Sleeping/Resting/Relaxing," tend to give people less happiness than people expect.

Activity	Predicted Happiness Rank of Activity	Actual Happiness Rank of Activity	Difference
Intimacy/Making Love	1	1	0
Pet Care/Playing with Pets	2	15	−13
Hobbies/Arts/Crafts	3	11	−8
Talking/Chatting/Socializing	4	7	−3
Theater/Dance/Concert	5	2	3
Singing/Performing	6	6	0
Sleeping/Resting/Relaxing	7	27	−20
Match/Sporting Event	8	13	−5
Computer Games/iPhone Games	9	20	−11
Watching TV/Film	10	19	−9
Birdwatching/Nature Watching	11	8	3
Eating/Snacking	12	21	−9
Other Games/Puzzles	13	16	−3
Hunting/Fishing	14	9	5
Gardening	15	5	10
Sports/Running/Exercise	16	4	12
Childcare/Playing with Children	17	14	3
Meditating/Religious Activities	18	12	6
Reading	19	24	−5
Exhibition/Museum/Library	20	3	17
Drinking Tea/Coffee	21	23	−2
Browsing the Internet	22	29	−7
Drinking Alcohol	23	10	13
Cooking/Preparing Food	24	22	2
Texting/Email/Social Media	25	30	−5
Listening to Speech/Podcast	26	25	1
Gambling/Betting	27	18	9
Traveling/Commuting	28	32	−4
Shopping/Errands	29	17	12
Care or Help for Adults	30	36	−6
Washing/Dressing/Grooming	31	26	5
Smoking	32	28	4
Working/Studying	33	37	−4
In a Meeting, Seminar, Class	34	33	1
Admin/Finances/Organizing	35	34	1
Housework/Chores/DIY	36	31	5
Waiting/Queueing	37	35	2
Sick in Bed	38	38	0

NOTES

INTRODUCTION: SELF-HELP FOR DATA GEEKS

2 *unconventional daters' success:* Christian Rudder, *Dataclysm: Who We Are (When We Think No One's Looking)* (New York: Broadway Books, 2014).

4 *career trajectories of hundreds of thousands of painters:* Samuel P. Fraiberger et al., "Quantifying reputation and success in art," *Science* 362(6416) (2018): 825–29.

5 *Oakland A's:* Michael Lewis, *Moneyball: The Art of Winning an Unfair Game* (New York: Norton, 2004).

6 *"more Moneyball than the Moneyball A's themselves":* Jared Diamond, "How to succeed in baseball without spending money," *Wall Street Journal*, October 1, 2019.

6 *track the trajectory of every shot:* Ben Dowsett, "How shot-tracking is changing the way basketball players fix their game," *FiveThirtyEight*, August 16, 2021, https://fivethirtyeight.com/features/how-shot -tracking-is-changing-the-way-basketball-players-fix -their-game/.

6 *forty-one shades of blue:* Douglas Bowman, "Goodbye, Google," https://stopdesign.com/archive/2009/03/20/goodbye-google.html, March 20, 2009.

6 *$200 million per year in additional ad revenue:* Alex Horn, "Why Google has 200m reasons to put engineers over designers," *Guardian*, February 5, 2014.

6 *"All others have to bring data":* "Are we better off with the internet?" YouTube, uploaded by the Aspen Institute, July 1, 2012, https://www.youtube.com/watch?v=djVrLNaFvIo.

7 *Renaissance's flagship Medallion fund:* Gregory Zuckerman, *The Man Who Solved the Market* (New York: Penguin, 2019).

7 *"they're just better than the rest of us":* Amy Whyte, "Famed Medallion fund 'stretches . . . explanation to the limit,' professor claims," *Institutional Investor*, January 26, 2020, https://www.institutionalinvestor.com/article/b1k2fymby99nj0/Famed-Medallion-Fund-Stretches-Explanation-to-the-Limit-Professor-Claims.

8 *unprecedented happiness dataset:* More details about Mappiness can be found at http://www.mappiness.org.uk.

10 *shifts are justified:* Rob Arthur and Ben Lindbergh, "Yes, the infield shift works. Probably," June 30, 2016, https://fivethirtyeight.com/features/yes-the-infield-shift-works-probably/.

10 *"we're all in sales now":* Daniel H. Pink, *To Sell Is Human* (New York: Penguin, 2012).

11 *having a poker face instead of a smile:* Neeraj Bharadwaj et al., "EXPRESS: A New Livestream Retail Analytics Framework to Assess the Sales Impact of Emotional Displays," *Journal of Marketing*, September 30, 2021.

14 *men sometimes type into Google:* Data on self-reported
    -in-Google-searches penis sizes can be found here:
    https://trends.google.com/trends/explore?date=all&q
    =my%20penis%20is%205%20inches,my%20
    penis%20is%204%20inches,my%20penis%20
    is%203%20inches,my%20penis%20is%206%20
    inches,my%20penis%20is%207%20inches.

15 *really like sentences that include the word "you":* Ariana
    Orwell, Ethan Kross, and Susan A. Gelman, " 'You'
    speaks to me: Effects of generic-you in creating
    resonance between people and ideas," *PNAS* 117(49)
    (2020): 31038–45.

15 *the best-selling books of all time:* https://en.wikipedia
    .org/wiki/List_of_best-selling_books.

16 *the most comprehensive study of rich people:* Matthew
    Smith, Danny Yagan, Owen Zidar, and Eric Zwick,
    "Capitalists in the Twenty-First Century," *Quarterly
    Journal of Economics* 134(4) (2019): 1675–1745.

16 *median age of entrepreneurs:* Pierre Azoulay, Benjamin
    F. Jones, J. Daniel Kim, and Javier Miranda, "Age and
    High-Growth Entrepreneurship," *American Economic
    Review* 2(1) (2020): 65–82.

17 *average successful entrepreneur is forty-two years old:*
    Ibid.

17 *the odds of starting a successful business increase:* Ibid.

17 *the advantage of age in entrepreneurship is true even in
    tech:* Ibid.

17 *Dataism:* Yuval Noah Harari, *Homo Deus: A Brief
    History of Tomorrow* (New York: Random House,
    2016).

18   *"organisms are algorithms"*: "Yuval Noah Harari. Organisms Are Algorithms. Body Is Calculator. Answer = Sensation~Feeling~Vedan?," YouTube, uploaded by Rashid Kapadia, June 13, 2020, https://www.youtube .com/watch?v=GrQ7nY-vevY.

18   *riddled with biases:* Daniel Kahneman, *Thinking, Fast and Slow* (New York: Farrar, Straus & Giroux, 2011).

CHAPTER 1: THE AI MARRIAGE

21   *"the most important decision that you make"*: https:// www.wesmoss.com/news/why-who-you-marry-is-the -most-important-decision-you-make/.

21   *compared the field of relationship science to an adolescent:* Harry T. Reis, "Steps toward the ripening of relationship science," *Personal Relationships* 14 (2007): 1–23.

22   *Joel's plan worked:* Samantha Joel et al., "Machine learning uncovers the most robust self-report predictors of relationship quality across 43 longitudinal couples studies," *PNAS* 117(32): 19061–71.

22   *The researchers had data on:* The variables examined can be found here: https://osf.io/8fzku/. The relevant file is Master Codebook With Theoretical Categorization, Final.xlsx, which is found in the section "Master Codebook with Theoretical Categorization." I thank Joel for pointing me to this file.

23   *Joel scheduled a talk in October 2019:* https://www .psychology.uwo.ca/pdfs/cvs/Joel.pdf.

24   *Samantha Joel told me in a Zoom interview:* I interviewed Joel on Zoom on September 24, 2020.

24 *reliably predict social unrest five days before it happens:* Ed Newton-Rex, "59 impressive things artificial intelligence can do today," *Business Insider,* May 7, 2017, https://www.businessinsider.com/artificial-intelligence-ai-most-impressive-achievements-2017-3#security-5.

24 *inform people of an emerging health issue:* Bernard Marr, "13 mind-blowing things artificial intelligence can already do today," *Forbes,* November 11, 2019, https://www.forbes.com/sites/bernardmarr/2019/11/11/13-mind-blowing-things-artificial-intelligence-can-already-do-today/#4736a3c76502.

26 *It is even possible to predict:* Jon Levy, David Markell, and Moran Cerf, "Polar Similars: Using massive mobile dating data to predict synchronization and similarity in dating preferences," *Frontiers in Psychology* 10 (2019).

27 *"unsolicited pictures of men's anatomy":* "What are single women's biggest complaints about online dating sites?" *Quora,* https://www.quora.com/What-are-single-womens-biggest-complaints-about-online-dating-sites; https://www.quora.com/What-disappointments-do-men-have-with-online-dating-sites.

28 *rate the importance of twenty-one traits in a potential romantic partner:* Harold T. Christensen, "Student views on mate selection," *Marriage and Family Living* 9(4) (1947): 85–88.

29 *The researchers found that looks matter:* Günter J. Hitsch, Ali Hortaçsu, and Dan Ariely, "What makes you click?—Mate preferences in online dating," *Quantitative Marketing and Economics* 8(4) (2010): 393–427. See Table 5.2.

31 *how height affected daters' desirability:* Ibid.

32 *Someone of a Desired Race:* https://www.gwern.net
    /docs/psychology/okcupid/howyourraceaffects
    themessagesyouget.html.

36 *effects of wealth on romantic desirability:* Hitsch,
    Hortaçsu, and Ariely, "What makes you click?"

37 *One's job matters in the mating market:* Ibid.

39 *sexiest names:* The results of this study were discussed in
    Daily Mail Reporter, "Why Kevins don't get girlfriends:
    Potential partners less likely to click on 'unattractive
    names' on dating websites," DailyMail.com, January 2,
    2012, https://www.dailymail.co.uk/news/article-2081
    166/Potential-partners-likely-click-unattractive
    -names-dating-websites.html. The academic study
    is Jochen E. Gebauer, Mark R. Leary, and Wiebke
    Neberich, "Unfortunate first names: Effects of name-
    based relational devaluation and interpersonal neglect,"
    Social Psychological and Personality Science 3(5) (2012):
    590–96.

39 *similarity, rather than difference, leads to attraction:*
    Emma Pierson, "In the end, people may really just
    want to date themselves," *FiveThirtyEight*, April 9,
    2014, https://fivethirtyeight.com/features/in-the
    -end-people-may-really-just-want-to-date-
    themselves/.

40 *Hinge users are 11.3 percent more likely:* Levy, Markell,
    and Cerf, "Polar Similars."

44 *Irrelevent Eight:* The success rate of different variables
    in predicting relationship happiness can be found in
    Tables 3, S4, and S5 of Joel et al., (2020).

47 *"He looked chubby in a uniform":* Alex Speier, "The transformation of Kevin Youkilis," *WEEI*, March 18, 2009.

49 *rate the attractiveness of each of their opposite-sex classmates:* Paul W. Eastwick and Lucy L. Hunt, "Relational mate value: consensus and uniqueness in romantic evaluations," *Journal of Personality and Social Psychology* 106(5) (2014): 728.

CHAPTER 2: LOCATION. LOCATION. LOCATION.

57 *in the first year of a baby's life:* Nehal Aggarwal, "Parents make 1,750 tough decisions in baby's first year, survey says," *The Bump*, July 9, 2020, https://www.thebump.com/news/tough-parenting-decisions-first-year-baby-life.

57 *parents have ranked the age of eight:* Allison Sadlier, "Americans with kids say this is the most difficult age to parent," *New York Post*, April 7, 2020.

58 *"Try timeouts":* Jessica Grose, "How to discipline without yelling or spanking," *New York Times*, April 2, 2019.

58 *"never use timeouts":* Wendy Thomas Russell, "Column: Why you should never use timeouts on your kids," *PBS NewsHour*, April 28, 2016.

58 *frustrated mother:* Rebecca Dube, "Exhausted new mom's hilarious take on 'expert' sleep advice goes viral," *Today*, April 23, 2013, https://www.today.com/moms/exhausted-new-moms-hilarious-take-expert-sleep-advice-goes-viral-6C9559908.

59 *three different worlds:* All median salaries are from the Bureau of Labor Statistics' Occupational Outlook Handbook, available at https://www.bls.gov/ooh/.

62 *counterexamples to the Emanuel lesson:* "I want to enroll a boy in dance class (ballet, etc.) but I fear he could be bullied because it's a 'girl thing' and also that he might become gay. What should I do?," *Quora*, https://www.quora.com/I-want-to-enroll-a-boy-in-dance-class-ballet-etc-but-I-fear-he-could-be-bullied-because-its-a-%E2%80%9Cgirl-thing-and-also-that-he-might-become-gay-What-should-I-do.

63 *Jim Lewis and Jim Springer:* The story of Jim Springer and Jim Lewis is discussed in many places, including in Edwin Chen, "Twins reared apart: A living lab," *New York Times*, December 9, 1979.

64 *"I've swung way over to the nature side":* Steve Lohr, "Creating Jobs: Apple's founder goes home again," *New York Times Magazine*, January 12, 1997.

65 *Harry and Bertha Holt:* The Holt story is told here: https://www.holtinternational.org/pas/adoptee/korea-2-adoptees/background-historical-information-korea-all/.

66 *Sacerdote's study:* Bruce Sacerdote, "How large are the effects from changes in family environment? A study of Korean American adoptees," *The Quarterly Journal of Economics* 122(1) (2007): 119–57.

69 *gave Jared a stake in his lucrative real estate business:* Andrew Prokop, "As Trump takes aim at affirmative action, let's remember how Jared Kushner got into Harvard," *Vox*, July 6, 2018, https://www.vox.com/policy-and-politics/2017/8/2/16084226/jared-kushner-harvard-affirmative-action.

69 *randomized controlled trial on breastfeeding:* Michael S. Kramer et al., "Effects of prolonged and exclusive breastfeeding on child height, weight, adiposity, and blood pressure at age 6.5 y: Evidence from a large randomized trial," *American Journal of Clinical Nutrition* 86(6) (2007): 1717–21.

69 *exposure to TV had no long-term effects on child test scores:* Matthew Gentzkow and Jesse M. Shapiro, "Preschool television viewing and adolescent test scores: Historical evidence from the Coleman Study," *Quarterly Journal of Economics* 123(1) (2008): 279–323.

69 *teaching kids cognitively demanding games:* John Jerrim et al., "Does teaching children how to play cognitively demanding games improve their educational attainment? Evidence from a randomized controlled trial of chess instruction in England," *Journal of Human Resources* 53(4) (2018): 993–1021.

70 *careful meta-analysis of bilingual education:* Hilde Lowell Gunnerud et al., "Is bilingualism related to a cognitive advantage in children? A systematic review and meta-analysis," *Psychological Bulletin* 146(12) (2020): 1059.

70 *"limited evidence":* Jan Burkhardt and Cathy Brennan, "The effects of recreational dance interventions on the health and well-being of children and young people: A systematic review," *Arts & Health* 4(2) (2012): 148–61.

72 *"It takes a family to raise a child":* "Acceptance Speech | Senator Bob Dole | 1996 Republican National Convention," YouTube, uploaded by Republican National Convention, March 25, 2016, https://www .youtube.com/watch?v=rYft9qxoLSo.

73   *some neighborhoods produce more successful kids:* Seth
Stephens-Davidowitz, "The geography of fame," *New
York Times,* March 13, 2014.

76   *SuperMetros:* Data on the causal effects of growing up
in different metropolitan areas can be found at http://
www.equality-of-opportunity.org/neighborhoods/.

77   *Within a given metropolitan area:* Raj Chetty et al.,
"The Opportunity Atlas: Mapping the childhood roots
of social mobility," NBER Working Paper 25147,
October 2018.

77   *some 13 percent:* In a section of the paper, the authors
say that a one standard deviation increase in mean
income in a Census tract is 21 percent of mean
household income; and that 62 percent of this effect is
due to causal effects of a neighborhood.

79   *some 25 percent:* If the standard deviation of the
income effects of parents overall is twice the standard
deviation of the income effects of a neighborhood,
the variance of the income effects of parents overall
is four times the variance of the income effects of a
neighborhood.

80   *most predictive of giving kids a big boost in life:* The
Tract-Level Correlations Between Neighborhoods,
Characteristics and Upward Mobility can be found
in Figures V and Figure II of the Online Appendix
in Chetty et al. (2018). These do not include data
for Student Teacher Ratio or School Expenditures.
Such data was studied in county-level estimates in
Raj Chetty and Nathaniel Hendren, "The impacts

of neighborhoods on intergenerational mobility II: county-level estimates," *Quarterly Journal of Economics* 133(3): 1163–28. The estimates can be found in Tables A.12 and Table A.14.

82 *The Power of Female Inventor Role Models:* Alex Bell et al., "Who becomes an inventor in America? The importance of exposure to innovation," *Quarterly Journal of Economics* 134(2) (2019): 647–713.

83 *The Power of Black Male Role Models:* Raj Chetty et al., "Race and economic opportunity in the United States: An intergenerational perspective," *Quarterly Journal of Economics* 135(2) (2019): 711–83.

CHAPTER 3: THE LIKELIEST PATH TO ATHLETIC GREATNESS IF YOU HAVE NO TALENT

93 *Phelps's short legs:* David Epstein, "Are athletes really getting faster, better, stronger?" TED2014, https://www.ted.com/talks/david_epstein_are_athletes_really_getting_faster_better_stronger/transcript?language=en#t-603684.

95 *Fence Your Way to College:* The O'Rourke story was told in Jason Notte, "Here are the best sports for a college scholarship," *Marketwatch.com*, November 7, 2018.

99 *4 times in every 1,000 pregnancies:* Christiaan Monden et al., "Twin Peaks: more twinning in humans than ever before," *Human Reproduction* 36(6) (2021): 1666–73.

100 *Twinsburg:* The Twinsburg festival has been discussed in many places, including Brandon Griggs, "Seeing double for science," *CNN*, August 2017.

101   *trusting behavior is 10 percent nature:* David Cesarini et al., "Heritability of cooperative behavior in the trust came," *PNAS* 105(10) (2008): 3721–26.

102   *ability to recognize sour tastes is 53 percent nature:* Paul M. Wise et al., "Twin study of the heritability of recognition thresholds for sour and salty tastes," *Chemical Senses* 32(8) (2007): 749–54.

102   *bullying is explained 61 percent by nature:* Harriet A. Ball et al., "Genetic and environmental influences on victims, bullies and bully-victims in childhood," *Journal of Child Psychology and Psychiatry* 49(1) (2008): 104–12.

102   *the T allele at rs11126630:* Irene Pappa et al., "A genome-wide approach to children's aggressive behavior," *American Journal of Medical Genetics* 171(5) (2016): 562–72.

103   *nine of them have been identical:* Estimates of which twins were identical came from news articles. There is conflicting information about whether Stephen and Joey Graham are identical or fraternal. There also wasn't information on whether Carl and Charles Thomas were identical or fraternal. So I reached out to Charles on LinkedIn. He responded and told me they are identical. Thanks, Charles!

103   *1 in 33,000:* This number of course depends on the year. But it can be calculated by comparing the total number of births in the United States in a given year to the total number of NBA players born in America in that year. For example, in 1990, there were roughly 4.2 million births in America, roughly half of whom were males. There have been 64 NBA players who were born in America in 1990.

103   *rough model based on the twins equations:* The code is on my website, sethsd.com, under the section "Twins Simulation Model."

104   *the difficulty scouts have evaluating identical twins:* Jeremy Woo, "The NBA draft guidelines for scouting twins," *Sports Illustrated*, March 21, 2018.

106   *6,778 Olympic wrestling athletes:* All estimated numbers of Olympic athletes are from Wikipedia.

CHAPTER 4: WHO IS SECRETLY RICH IN AMERICA?

113   *Jack MacDonald:* Katherine Long, "Seattle man's frugal life leaves rich legacy for 3 institutions," *Seattle Times*, November 26, 2013.

113   *Anna Sorokin:* Rachel Deloache Williams, "My bright-lights misadventure with a magician of Manhattan," *Vanity Fair*, April 13, 2018.

115   *paid $8.9 million in 2019 by Stanford University:* Steve Berkowitz, "Stanford football coach David Shaw credited with more than $8.9 million in pay for 2019," *USA Today*, August 4, 2021.

116   *Jerry Richardson:* Nick Maggiulli (@dollarsanddata), "2. Not thinking like an owner Do you know who the wealthiest NFL player in history is? Not Brady/Manning/Madden. It's Jerry Richardson. Never heard of him? Me neither. He made his wealth from owning Hardees franchises, not playing in the NFL. Be an owner. Think like one too." February 8, 2021, 12:30 P.M., tweet.

117   *which business fields give an owner:* Tian Luo and Philip B. Stark, "Only the bad die young: Restaurant mortality in the Western US," arXiv: 1410.8603, October 31, 2014.

119   *Top 5 Businesses with the Greatest Number of Millionaires:* This is from the online appendix of Smith, Yagan, Zidar, and Zwick, "Capitalists in the Twenty-First Century." In particular, the data comes from Table J.3 in http://www.ericzwick.com/capitalists /capitalists_appendix.pdf. I thank Eric Zwick for pointing me to it.

124   *are at least 10,000 independent creatives:* This includes rich owners of S-Corporations and partnerships.

CHAPTER 5: THE LONG, BORING SLOG OF SUCCESS

137   *Tony Fadell:* Fadell's story was told in many places, including here: Seema Jayachandran, "Founders of successful tech companies are mostly middle-aged," *New York Times,* September 1, 2019.

138   *In an interview on* The Tim Ferriss Show, *Fadell said: The Tim Ferriss Show* #403, "Tony Fadell— On Building the iPod, iPhone, Nest, and a Life of Curiosity," December 23, 2019.

140   *"people over 45 basically die in terms of new ideas":* Corinne Purtill, "The success of whiz kid entrepreneurs is a myth," *Quartz,* April 24, 2018.

140   *"Young people are just smarter":* Lawrence R. Samuel, "Young people are just smarter," *Psychology Today,* October 2, 2017.

142   *years following the release of the movie:* "Surge in teenagers setting up businesses, study suggests," https:// www.bbc.com/news/newsbeat-50938854.

143 *Suzy Batiz:* Carina Chocano, "Suzy Batiz' empire of odor," *New Yorker*, November 4, 2019; Liz McNeil, "How Poo-Pourri founder Suzy Batiz turned stinky bathrooms into a $240 million empire," *People*, July 9, 2020.

144 *"The Outsider Advantage":* David J. Epstein, *Range* (New York: Penguin, 2019).

146 *"great new things often come from the margins":* Paul Graham, "The power of the marginal," paulgraham .com, http://www.paulgraham.com/marginal.html.

150 *NBA players disproportionately come from middle-class families:* Joshua Kjerulf Dubrow and Jimi Adams, "Hoop inequalities: Race, class and family structure background and the odds of playing in the National Basketball Association," *International Review for the Sociology of Sport* 45(3): 251–57; Seth Stephens-Davidowitz, "In the N.B.A., ZIP code matters," *New York Times,* November 3, 2013.

151 *people are more likely to laugh when things are going well:* Seth Stephens-Davidowitz, "Why are you laughing?" *New York Times,* May 15, 2016.

151 *more intelligence is always an advantage:* Matt Brown, Jonathan Wai, and Christopher Chabris, "Can you ever be too smart for your own good? Comparing linear and nonlinear effects of cognitive ability on life outcomes," PsyArXiv Preprints, January 30, 2020.

CHAPTER 6: HACKING LUCK TO YOUR ADVANTAGE

157 *Airbnb:* The Airbnb story has been told many places, including by Leigh Gallagher, *The Airbnb Story: How*

Three Ordinary Guys Disrupted an Industry, Made Billions . . . and Created Plenty of Controversy (New York: HMH Books, 2017).

159 *"Idea times Product times Execution times Team times Luck":* Tad Friend, "Sam Altman's manifest destiny," *New Yorker*, October 3, 2016.

160 *how luck influences large companies:* Jim Collins, *Great by Choice (Good to Great)* (New York: Harper Business, 2011).

163 *Airbnb's bookings dropped:* Corrie Driebusch, Maureen Farrell, and Cara Lombardo, "Airbnb plans to file for IPO in August," *Wall Street Journal*, August 12, 2020.

163 *IPO'd at a more than $100 billion valuation:* Bobby Allyn and Avie Schneider, "Airbnb now a $100 Billion company after stock market debut sees stock price double," *NPR*, December 10, 2020.

164 *a clear difference between sports and art:* Albert-László Barabási, *The Formula* (New York: Little, Brown, 2018).

165 *Only 7 of the 1,097 people who passed by stopped to listen:* Gene Weingarten, "Pearls Before Breakfast: Can one of the nation's great musicians cut through the fog of a D.C. rush hour? Let's find out," *Washington Post*, April 8, 2007.

166 *the* Mona Lisa *was gone:* R. A. Scotti, *Vanished Smile* (New York: Vintage, 2009).

168 *"Da Vinci Effect":* https://www.beervanablog.com /beervana/2017/11/16/the-da-vinci-effect.

168 Salvator Mundi: Caryn James, "Where is the world's most expensive painting?," BBC.com, August 19, 2021,

https://www.bbc.com/culture/article/20210819-where
-is-the-worlds-most-expensive-painting.

169   *what predicts success in the art world:* Fraiberger et al.,
"Quantifying reputation and success in art."

171   *schedule of a Category 1 Artist as a Young Man:* Painting
schedules of particular artists from the dataset were
kindly provided by Samuel P. Fraiberger.

174   *how Springsteen diagnosed his problem:* "The Promised
Land (Introduction Part 1) (Springsteen on Broadway -
Official Audio)," YouTube, uploaded by Bruce
Springsteen, December 14, 2018, https://www.you
tube.com/watch?v=omuusrmb6jo&list=PL9tY0BWXO
ZFs9l_PMss5AB8SD38lFBLwp&index=12.

177   *Artists who produce more work tend to have more hits:*
Dean Keith Simonton, "Creativity as blind variation
and selective retention: Is the creative process
Darwinian?," *Psychological Inquiry* 10 (1999): 309–28.

177   *"You wrote it": No Direction Home*, directed by Martin
Scorsese, Paramount Pictures, 2005.

178   *Beethoven disliked a piece he had created:* Aaron Kozbelt,
"A quantitative analysis of Beethoven as self-critic:
Implications for psychological theories of musical
creativity," *Psychology of Music* 35 (2007): 144–68.

178   *Upon completing his third album,* Born to Run*:* Louis
Masur, *"Tramps Like Us:* The birth of *Born to Run*,"
Slate, September 2009, https://slate.com/culture
/2009/09/born-to-run-the-groundbreaking-springsteen
-album-almost-didnt-get-released.html.

181   *When the least desirable men on the site:* Elizabeth E.
Bruch and M. E. J. Newman, "Aspirational pursuit of

mates in online dating markets," *Science Advances* 4(8) (2018).

182   *heterosexual women significantly increase:* Derek A. Kraeger et al., " 'Where have all the good men gone?' Gendered interactions in online dating," *Journal of Marriage and Family* 76(2) (2014): 387–410.

185   *Chris McKinlay:* Kevin Poulsen, "How a math genius hacked OkCupid to find true love," *Wired*, January 21, 2014. McKinlay tells the story in his book *Optimal Cupid: Mastering the Hidden Logic of OkCupid* (CreateSpace Independent Publishing Platform, 2014).

186   *for every offer a scientist receives:* Jason D. Fernandes et al., "Research culture: A survey-based analysis of the academic job market," *eLife Sciences*, June 12, 2020.

CHAPTER 7: MAKEOVER: NERD EDITION

190   *the world's foremost expert on faces:* Alexander Todorov, *Face Value* (Princeton, NJ: Princeton University Press, 2017). I also interviewed Todorov on May 7, 2019.

194   *the person whose face was judged as more competent:* Alexander Todorov et al., "Inferences of competence from faces predict election outcomes," *Science* 308(5728) (2005): 1623–26.

195   *what best predicted which West Point cadets:* Ulrich Mueller and Allan Mazur, "Facial dominance of West Point cadets as a predictor of later military rank," *Social Forces* 74(3) (1996): 823–50.

198   *rate multiple pictures of the same person on many dimensions:* Alexander Todorov and Jenny M. Porter, "Misleading first impressions: Different for different facial images of the same person," *Psychological Science* 25(7) (2014): 1404–17.

CHAPTER 8: THE LIFE-CHANGING MAGIC OF LEAVING YOUR COUCH

210 *tenure does not give lasting happiness:* Dan Gilbert et al., "Immune neglect: A source of durability bias in affective forecasting," *Journal of Personality and Social Psychology* 75(3) (1998): 617–38.

210 *what it is like being denied tenure:* "What is it like to be denied tenure as a professor?," *Quora,* https://www.quora.com/What-is-it-like-to-be-denied-tenure-as-a-professor.

213 *recruited a whole bunch of colonoscopy patients:* Donald A. Redelmeier and Daniel Kahneman, "Patients' memories of painful medical treatments: Real-time and retrospective evaluations of two minimally invasive procedures," *Pain* 66(1) (1996): 3–8.

CHAPTER 9: THE MISERY-INDUCING TRAPS OF MODERN LIFE

234 *search engines are worth $17,530 every year to the average American:* Erik Brynjolfsson, Avinash Collis, and Felix Eggers, "Using massive online choice experiments to measure changes in well-being," *PNAS* 116(15) (2019): 7250–55.

234 *30 percent of Americans said they were "very happy":* GSS data is available here: https://gssdataexplorer.norc.org/trends/Gender%20&%20Marriage?measure=happy.

235   *income does increase happiness, but the effects are small:*
Matthew A. Killingsworth, "Experienced well-being
rises with income, even above $75,000 per year," *PNAS*
118(4) (2021).

236   *meditation does boost happiness:* Xianglong Zeng
et al., "The effect of loving-kindness meditation on
positive emotions: A meta-analytic review," *Frontiers in
Psychology* 6 (2015): 1693.

238   *Work sucks:* Alex Bryson and George MacKerron,
"Are you happy while you work?" *Economic Journal*
127(599) (2016): 106–25.

245   *experiment on the effects of using Facebook:* Hunt Allcott
et al., "The welfare effects of social media," *American
Economic Review* 110(3) (2020): 629–76.

247   *how sports fans' happiness was affected:* Peter Dolton
and George MacKerron, "Is football a matter of life or
death—or is it more important than that?," National
Institute of Economic and Social Research Discussion
Papers 493, 2018.

249   *"Better than the Knicks":* Sean Deveney, "Andrew Yang
brings his hoop game, 2020 campaign to A.N.H. gym
for new series," https://www.forbes.com/sites/seand
eveney/2019/10/14/andrew-yang-2020-campaign-new
-hampshire-luke-bonner/?sh=73927bbf1e47.

251   *"I don't like drinking":* "Comedians Tackling
Depression & Anxiety Makes Us Feel Seen," YouTube,
uploaded by Participant, https://www.youtube.com
/watch?v=TBV-7_qGlr4&t=691s.

252   *happiness of booze:* Ben Baumberg Geiger and
George MacKerron, "Can alcohol make you happy?

A subjective wellbeing approach," *Social Science & Medicine* 156 (2016): 184–91.

255   *being in nature is a crucial component of happiness:* George MacKerron and Susana Mourato, "Happiness is greater in natural environments," *Global Environmental Change* 23(5) (2013): 992–1000.

259   *being around beauty itself:* Chanuki Illushka Seresinhe et al., "Happiness is greater in more scenic locations," *Scientific Reports* 9 (2019): 4498.

261   *HappyHier:* Sjerp de Vries et al., "In which natural environments are people happiest? Large-scale experience sampling in the Netherlands," *Landscape and Urban Planning* 205 (2021).

262   *frequently, other factors matter more than the weather:* All happiness comparisons are the author's calculations based on Table 2 here: https://eprints.lse.ac.uk/49376 /1/Mourato_Happiness_greater_natural_2013.pdf.

INDEX

ABOUT THE AUTHOR

Seth Stephens-Davidowitz is a data scientist, author, and keynote speaker. His 2017 book, *Everybody Lies*, was a *New York Times* bestseller and an *Economist* Book of the Year. He has worked as a contributing op-ed writer for the *New York Times*, a lecturer at the Wharton School, and a Google data scientist. He received a BA in philosophy from Stanford, where he graduated Phi Beta Kappa, and a PhD in economics from Harvard. He lives in Brooklyn and is a passionate fan of the Mets, Knicks, Jets, and Leonard Cohen.